Laboratory
Experiments
for
Basic
Chemistry

Laboratory
Experiments
for
Basic
Chemistry
Fourth Edition

Alan Sherman
Sharon Sherman
Leonard Russikoff

Middlesex County College
Edison, New Jersey

Houghton Mifflin Company   Boston
Dallas   Geneva, Illinois   Palo Alto   Princeton, New Jersey

Cover photograph by Ralph Mercer.

Figures 11-1 and 23-1, pages 97 and 197, are used by permission of Bausch & Lomb. Table 1, page 224, is used by permission of the Chemical Rubber Company.

ISBN: 0-395-44877-8

ABCDEFGHIJ-B-9543210/8987

Contents

Preface

Laboratory Experiments for Basic Chemistry was written for use in one-term or two-quarter introductory chemistry courses for students with a very limited background in science and mathematics. The experiments provide support for key principles covered in lectures and give students an opportunity to learn skills that will be useful in the further study of college chemistry. The manual is organized so that it follows our textbook, *Basic Concepts of Chemistry.* However, we have included sufficient additional experiments, some on subjects not covered in the textbook, to enable an instructor to adopt it as a laboratory manual to accompany other textbooks written for courses in preparatory chemistry and chemistry for liberal arts majors.

This manual addresses a complete range of objectives, which are clearly stated with each experiment. Such basic lab skills as classification, weighing, measuring, and separating various materials are covered in the initial experiments. Later experiments deal with basic concepts of chemistry, focusing on the difference between elements, compounds, and mixtures, and the chemistry of elements within a group.

Beginning with the second edition, experiments on Carbohydrates, Lipids, and Proteins, the Determination of the Amount of Iron in Water, and the Determination of the Dissolved Oxygen Content of Water were added. These new experiments introduced students to three classes of biochemical compounds, gave instruction in conducting a colorimetric test on potable water, and taught the basics of performing oxidation-reduction titrations. The second edition also added an exercise called CHEM-DECK© , a card game that was devised to teach students how to write chemical compounds while playing CHEM-DECK© Poker or CHEM-DECK© Gin Rummy. In the third edition an experiment on Chemical Kinetics was added to coincide with the Kinetics and Equilibrium chapter in the text.

New to the fourth edition is a Chemical Safety and Hazard Guide. In several states, environmental legislation that has either been passed or is pending mandates that individuals working with chemicals be given access to information about these chemicals. Our Chemical Safety and Hazard Guide, which appears at the beginning of each experiment, fulfills this mandate. It is also a way of further emphasizing to students the safe handling of chemicals in the laboratory. In addition, minor adjustments were made in several experiments with an eye towards safety. For example, chemicals such as benzene and carbon tetrachloride are no longer used in any of the experiments.

In order to keep the manual flexible we have included several additional experiments for the instructor to choose from, realizing that in one semester an instructor could not cover all of the experiments offered. In effect, we are providing a variety of experiments to fit a wide range of needs.

The key features of the manual are as follows:

1. A list of "things you will learn by doing this experiment" precedes each experiment.

2. Each experiment begins with a complete list of materials the student needs in order to perform the experiment. We include chemicals that are relatively inexpensive and equipment that is commonly used. An alphabetical listing of materials needed in the experiments is given at the end of the manual.

3. We include discussions of the basic principles underlying each experiment. In some cases we give actual laboratory data that was gathered as the manual was class-tested. This enables the student to see exactly what is involved in performing each experiment.

4. We give step-by-step procedures in outline form. We warn against possible hazards and list precise safety rules. We also point out possible complications that could develop in some experiments.

5. Clear diagrams accompany each experiment, showing the design and setup.

6. Exclusive use of the metric notation is followed. We provide extensive coverage of English-metric conversions in our textbook.

7. The manual provides thought-provoking questions for class discussion and individual study.

8. We include report pages that are geared to help students record the data and understand the calculations involved in each experiment.

In addition, an Instructor's Manual (which is included in the Instructor's Manual for the textbook, *Basic Concepts of Chemistry,* 4th edition) is available. This Instructor's Manual gives suggestions and short cuts for managing these experiments efficiently and safely. The Instructor's Manual discusses each experiment in detail and gives directions for preparing the materials needed.

We have based this laboratory manual on the experience we have gained by teaching basic chemistry at Middlesex County College in Edison, New Jersey, since 1969. The manual constitutes a complete program of experiments and exercises. It does not omit basic skills or concepts in order to simplify the presentation.

We recommend doing experiments 1, 2, 3, and 4 in the order presented, so that students will gain proficiency in basic laboratory techniques first. After they do so, we invite users to choose from the remaining material and tailor the experiments to their particular courses.

Among the students who gain significantly from using this laboratory manual are those who have taken relatively few courses in science and mathematics. Often we find that students who benefit a great deal from our course had little or no success in their high school science and mathematics courses. Adults who are returning to school also do well with this manual. Some studied high school chemistry years ago, and at this point they recall very little of what they learned. Some may be wary of taking a science course. Those whose skills in science and mathematics require remediation generally find the material offered here quite appropriate. The simple, basic approach we use enables students to overcome their timidity and regain skills and confidence.

We would like to thank several former and current reviewers without whose help the present form of our laboratory manual would be neither as useful nor as complete. Larry K. Krannich of the University of Alabama at Birmingham deserves special thanks for suggesting many helpful points. So, too, do Arnold Loebel of Merritt College in Oakland, California; Stanley M. Cherim of Delaware County Community College in Pennsylvania; David G. Williamson of California Polytechnic State University at San Luis Obispo; Charles Oxman of Middlesex County College; and Wilma Meckstroth of the Ohio State University, Newark Campus.

We hope that you enjoy doing these experiments and invite you to send us any suggestions for improvements.

Alan Sherman
Sharon J. Sherman
Leonard J. Russikoff
Middlesex County College
Edison, NJ 08818

Safety Regulations for Chemistry Laboratories

Read these safety regulations carefully, and be sure you understand them. Before each laboratory session, your instructor will discuss any safety hazards that might be associated with that day's experiment.

1. Report all accidents to your instructor.

2. Wear safety goggles and a laboratory apron in the laboratory at all times. Always wear eye covering that will protect your eyes against both impact and splashes. (If you should get a chemical in your eye, wash the eye with flowing water from the sink or fountain for 15 to 20 minutes.)

3. Do not perform any unauthorized experiments.

4. In case of fire or accident, call the instructor at once. Note the location of fire extinguisher, safety shower, fire blanket, eyewash, and phone, so you can use them quickly in an emergency.

5. If you cut or burn yourself or accidently inhale fumes, notify your instructor at once. The instructor will arrange immediate treatment according to the regulations of your college.

6. Do not taste anything in the laboratory. (This applies to food as well as chemicals. Do not use the laboratory as an eating place; never eat or drink from laboratory glassware.)

7. Exercise great care in noting the odor of fumes and avoid breathing fumes of any kind.

8. Do not use mouth suction to fill pipettes with chemical reagents. (Use a suction bulb to fill pipettes.)

9. Do not force glass tubing into rubber stoppers. Lubricate the tubing and introduce it gradually and gently. Protect your hands with a towel when you are inserting lubricated tubing into stoppers.

10. Confine long hair whenever you are in the laboratory.

11. Place all hot glassware on a heat-resistant mat to cool; this will also signify to all laboratory personnel that the glassware is *hot*. Do not hand hot glassware to another person, because a person's natural instinct is to reach for it.

12. Corrosive acids and bases are very soluble in water. If either a corrosive acid or base comes in contact with your skin, you can wash it off your skin before much damage is done. Haste in washing the affected area is essential. Summon the laboratory instructor if you spill a corrosive acid or base on your skin.

13. Carry out experiments in which noxious fumes are produced, or in which there is danger of explosion, under a fume hood with the safety shield pulled down for protection.

14. Do not wear open-toed shoes or shorts in the laboratory, since they do not offer enough protection to the body.

15. *Never* point a test tube containing a reacting mixture (especially when you are heating it) toward another person or toward yourself.

16. If you are preparing a dilute acid solution, never pour water into concentrated acid. Always pour the acid into the water while stirring the water constantly.

17. Be extremely cautious when you are lighting a Bunsen burner.

18. Never engage in horseplay in the laboratory.

19. Read the label carefully before removing a chemical from its container.

20. Never work in the laboratory alone.

21. *No Smoking* in or around the laboratory area!

Please sign the form below and give it to your instructor.

I, the undersigned, have read the Safety Regulations for Chemistry Laboratories, and I understand them.

(Signature) _____

(Date) _____

Experiment 1
Classification

Time: About 2 hours

Materials: A selection of buttons (enough so that each group of students may have about 50), bar magnet, a selection of various chemical substances (labeled only by a letter code) in sealed test tubes or vials

. . . .

Safety and Chemical Hazard Guide

Please review this before performing the experiment.

Each of the experiments in this laboratory manual is preceded by a section entitled "Safety and Chemical Hazard Guide." It is important for you to review the information presented here, to become aware of the nature of the materials with which you will be working.

This experiment, Classification, has no specific safety precautions because the chemicals you will observe are sealed and you will not come into contact with any of them. However, please note that safety glasses are to be worn at all times in the laboratory.

INTRODUCTION

Putting objects into groups because they share some similar characteristic is called classification. Nearly all branches of science use systems of classification. Systems of classification are necessary to help the scientist handle the huge amounts of data and facts that have accumulated over the years. When individual objects are placed into groups based on properties they have in common, the study of a large number of objects is made much easier.

Some Things You Will Learn by Doing This Experiment

1. You will learn many ways to classify a group of objects. Each time you classify an object according to a different aspect of it, you learn more about it.

1

2. You will learn how classification of a large number of objects into a few categories makes it easier to study the objects, because all the objects in a specific group have something in common.

DISCUSSION

In this exercise you will have a chance to devise your own classification system. You will make your own rules for setting up categories and subcategories. To be a successful classifier, you will have to make use of all your powers of observation, especially your sense of sight. In other experiments, you'll make use of your other senses as well. Good luck!

PART 1: THE CLASSIFICATION OF BUTTONS

You will be given a handful of buttons. Buttons, of course, are common objects that you see every day. But have you ever stopped to think that buttons could be put into groups because they share certain similar characteristics? Examine the buttons and decide how you want to classify them. Some people form groups based on the sizes, shapes, colors, and other characteristics of the buttons. When you have decided how to group the buttons, you should then form subgroups and further subdivide, if possible. For example, if you choose to group the buttons by size, you can have large buttons, large round buttons, and large white round buttons. After you have classified them, make a chart showing how your classification scheme works. Also list the number of buttons in each group and subgroup (see the example below). If time permits, repeat the experiment using different criteria for the groups. Compare your classification system with those of other people in the class.

Sample Chart

Classification by Size

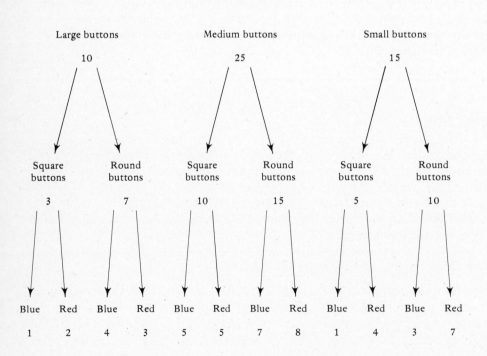

PART 2: THE CLASSIFICATION OF SOME CHEMICAL SUBSTANCES

. . . .

Note: Be sure you have your safety glasses on for this part of the experiment.

. . . .

In this part of the experiment you will try to classify some chemical substances. First get your set of substances from the instructor. Notice that the test tubes and vials are sealed. In this experiment you will not remove the contents of the test tubes and vials. Again, make your own rules for setting up groups and subgroups. You may want to use criteria such as the appearance (texture, color, and so forth) of each substance, or the physical state (solid, liquid, gas) of each substance. The instructor will place a bar magnet at your disposal if you wish to use it as an aid in deciding on a classification scheme. In a later experiment we'll look at some of these substances again and see how most chemists classify them.

Prepare a chart similar to the one you prepared in Part 1, to show the results of your classification scheme. If time permits, try a second classification. Compare your system with those of other people in the class.

. . . .

Caution: Be extremely careful with the vials containing the mercury, bromine, and chlorine. These materials are very hazardous, and problems can arise if the vials are broken.

. . . .

QUESTIONS

1. List some of the advantages of *your* classification system for buttons.
2. List some of the advantages of *your* classification system for the chemical substances.

©1988 Houghton Mifflin Company

Report on Experiment 1

Name _____

Section _____ Date _____

Instructor _____

PART 1: THE CLASSIFICATION OF BUTTONS

Use this report page to show your classification system.

PART 2: THE CLASSIFICATION OF SOME CHEMICAL SUBSTANCES

Use this report page to show your classification system.

Report on Experiment 1

Name _____

Section _____   Date _____

Instructor _____

RESPONSES TO QUESTIONS

1.

2.

Experiment 2
Use of the Balance: Determining the Densities of Some Common Objects

Time: About 2 hours

Materials: Triple-beam balance, aluminum weighing pans, beakers (50 mL, 100 mL), graduated cylinders (10 mL, 25 mL, 50 mL), metric rulers, objects of varying density (distilled water, a small block of wood, an iron cylinder, a glass marble, ethyl alcohol)

. . . .

Safety and Chemical Hazard Guide

Please review this before performing the experiment.

1. Glasses must be worn at *all* times when in the laboratory!
2. Metal cylinders (copper, aluminum, iron). No hazards.
3. Ethyl alcohol. Intoxicating, do not inhale. Ingestion of *large* amounts can cause nausea, vomiting, flushing, mental excitement or depression, drowsiness, impaired perception, incoordination, stupor, coma or death.

INTRODUCTION

One of the most important instruments in the laboratory is the balance. With a balance, the chemist is able to weigh materials with great accuracy and precision. There are many different types of balances available to the chemist today; Figure 2-1 shows some of them.

There are balances that weigh things with a precision as high as 0.0001 g (0.1 mg), and others weigh with a precision of only 0.1 g or 0.01 g. Some balances work automatically. The chemist simply puts the thing to be weighed on the balance pan, turns the balance to "on," and reads the weight of the thing directly. Other balances require the chemist to turn knobs or add weights in order to weigh an object. (Remember that, although we use the terms *weigh* and *weight*, what we're actually doing is determining the mass of objects. Our balances contain masses, not

Top-loading balance

Triple-beam balance

Triple-beam
balance

Figure 2-1
Various types of balances used in a chemical laboratory

weights. Mass is the amount of matter in an object, whereas weight is the gravitational force with which a planet attracts an object.*)

In this experiment you will learn how to use a balance. Once you have mastered the use of the balance, you will use this skill to find the densities of some common objects.

Some Things You Will Learn by Doing This Experiment

1. You will learn how to weigh objects on a triple-beam balance.
2. You will learn how to measure volumes with a graduated cylinder.
3. You will learn how to compute volume by physical measurement.
4. You will learn how to measure volume by water displacement.
5. You will learn how to figure out the densities of various substances.

DISCUSSION

PART 1: THE TRIPLE-BEAM BALANCE

The balance used in many chemistry laboratory courses in colleges today is the triple-beam balance. This balance is very useful, because you can weigh objects to a precision of 0.01 g (that's one one-hundredth of a gram). There are several dif-

*For further discussion of the differences between mass and weight, see Chapter 2 of our text, *Basic Concepts of Chemistry,* fourth edition (Boston: Houghton Mifflin, 1988).

ferent types of triple-beam balances on the market today. Therefore your instructor will give you detailed instructions in the operation of the one you will be using. However, there are some general rules that apply to the handling of *all* balances, and you must follow these rules carefully.

PROCEDURE

1. When you first look at your balance, note its weighing capacity. Do not place objects on the balance that you think might exceed this capacity.

2. Check to see whether the balance is in good working order. The pan should swing freely. If it does not, check to see whether your balance has a beam-release mechanism, which raises the beam off the knife edge when the balance is not in use. Gently turn the beam-release switch to place the balance in its working position.

3. Check the zero point of the balance. Be sure all weights are notched into their zero positions. Then check to see whether the pointer on the balance is at zero. If it is not, ask your instructor to help you adjust the balance.

4. Never place powder or crystalline chemicals directly on the metal (or plastic) weighing pan. Corrosive chemicals damage metal pans, and many chemicals permanently stain plastic pans. Be careful never to place hot objects on the weighing pans. Hot objects can melt plastic pans and can cause warping and scarring of metal pans.

5. To weigh powder or crystalline chemicals, use an aluminum weighing cup, a watch glass, or a piece of weighing paper (or any smooth-surface paper). Put the cup on the weighing pan and find the weight of the empty cup. Then put the chemical to be weighed in the cup. You can then find the weight of the cup plus the chemical. To figure out the weight of the chemical, all you do is subtract the cup weight from the total weight of cup plus chemical.

   Weight of chemical + cup
   — Weight of cup
   _____
   Weight of chemical

6. You can find the weight of a liquid in a similar manner. First weigh a graduated cylinder or a beaker when it is empty. Then put the liquid to be weighed into the graduated cylinder or beaker and determine the weight of the liquid by subtraction.

7. After weighing an object, remember to return all the weights to their zero positions.

8. Keep the balance dry. If you spill something on the balance pan, wipe it immediately. If the material is an acid or base, wipe down the area with a damp sponge before drying. Notify your instructor.

9. Make sure your hands are dry and clean before you touch the weights on the balance.

10. Always leave the balance clean and ready to be used for the next weighing.

PROCEDURE FOR WEIGHING EXERCISE

1. Be sure that you have read the previous section and that you have detailed instructions on the use of your balance.

. . . .

Note: Before you proceed any further, *check to make sure that you have your safety glasses on.* Make it a habit to put on your safety glasses the minute you walk into the lab.

. . . .

2. Obtain a 10-mL graduated cylinder and weigh it.
3. Record the weight of the graduated cylinder on the report page at the end of this experiment. Record all weights to *two* decimal places, unless your instructor gives you other directions.
4. Have your instructor check your reading to be sure you have weighed and recorded the graduated cylinder correctly.
5. Repeat steps 2 through 4, using the following items:
   a) a 25-mL graduated cylinder
   b) a 50-mL beaker
   c) a 100-mL beaker
   d) an aluminum weighing cup

PART 2: DETERMINING THE DENSITIES OF SOME COMMON OBJECTS

The density of an object is defined as its mass per unit volume.

Mathematically we can state that

$$\text{Density} = \frac{\text{mass}}{\text{volume}} \quad \text{or} \quad D = \frac{M}{V}$$

When we measure the mass of an object in grams and its volume in cubic centimeters (remember that a cubic centimeter, cm^3 or cc, is the same as a milliliter, mL), we can express the density of the object in terms of grams per cubic centimeter.

$$D = \frac{\text{g}}{\text{cc}}$$

In this experiment we shall determine the densities of some common objects. To do this properly, you will have to weigh each object and measure its volume. By dividing the weight of the object by its volume, you can compute its density. When reporting the density of each object, be sure to use the proper number of significant figures (Chapter 2 of the text has a section on how to use significant figures).

PROCEDURE FOR DETERMINING THE DENSITIES OF FIVE MATERIALS

a) DENSITY OF WATER

1. Weigh a clean, dry graduated cylinder.
2. Record its weight, to two decimal places, in the data table at the end of this experiment.
3. Add water to your graduated cylinder until you have approximately 10 mL of water. Read the volume of the water to the nearest 0.1 mL and record the volume on the report page.

Figure 2-2
Reading the volume of water in a graduated cylinder

. . . .

Note: When you're reading the volume of water in a graduated cylinder, read the bottom of the concave meniscus (Figure 2-2). Also be sure that the graduated cylinder is at eye level.

. . . .

4. Find the weight of the water plus the graduated cylinder and record it in the data table.

5. Find the weight of the water by subtraction and record this also.

6. Compute the density of the water by dividing the weight of the water by its volume. Report to the proper number of significant figures and write this in your data table.

b) DENSITY OF WOOD

1. Get a block of wood. Measure the length, width, and height in centimeters, using a metric ruler. You should be able to measure to 0.1 cm.

2. Compute the volume of the block by multiplying its length times its width times its height:

$V = l \times w \times h$

3. Write in the data table the volume of the block in cubic centimeters. Report this volume to the proper number of significant figures.

4. Weigh the block of wood (you may weigh it directly on the balance pan) and record the weight.

5. Compute the density of the wood and record this figure.

c) DENSITY OF IRON

1. Weigh the iron cylinder and record its weight on the report page.

2. We shall determine the volume of an iron cylinder in two ways, by direct measurement and by water displacement.

3. To find the volume of the iron cylinder by measurement, use a metric ruler and measure the height and then the diameter of the cylinder (Figure 2-3).

4. Compute the volume of the cylinder (in cubic centimeters) using the formula

$$V = \frac{\pi d^2 h}{4}$$

where d = diameter of the cylinder, h = height of the cylinder, and π = 3.14

Figure 2-3
Measuring the iron cylinder

5. To find the volume of the iron cylinder by water displacement, obtain a 50-mL graduated cylinder. Fill it to exactly 25.0 mL with water (this is your initial reading). Gently lower the iron cylinder into the water. Read the level to which the water rises in the graduated cylinder (this is your final reading). Then subtract this level from the original water level. Record all data on the report page. The difference in water levels is due to the displacement of water by the iron cylinder and represents the volume occupied by the iron cylinder. Compare this with the calculated volume. Are your measurements similar?

6. Compute the density of the iron cylinder, using first the calculated volume and then the water-displacement volume, and write these densities in the appropriate places on the report page.

d) DENSITY OF GLASS MARBLE

1. Weigh the glass marble and record its weight on the report page.
2. Find the volume of the marble by water displacement and record it on the report page.
3. Compute the density of the marble and record it also.

e) DENSITY OF ETHYL ALCOHOL

1. Weigh a clean, dry 25-mL graduated cylinder.
2. Pour exactly 20.0 mL of ethyl alcohol into the graduated cylinder. Record this volume on the report page.
3. Weigh the ethyl alcohol plus the graduated cylinder and record this weight.
4. Use subtraction to obtain the weight of the ethyl alcohol and record it.
5. Now compute the density of the ethyl alcohol and list it in your data table.
6. Using the *Handbook of Chemistry and Physics*, find the values it reports for the densities of ethyl alcohol and iron. List these values in the data table and compare them with your values.

QUESTIONS

1. You calculated densities in this experiment using objects at room temperature (about 25°C). How do you think the density of the ethyl alcohol would change if it were warmed to 50°C before you did the experiment? Would the density be greater or less than the density determined at 25°C? Explain.

Figure 2-4
The rock

2. How could you compute the density of the human body? If you were asked to guess the density of the human body, what number would you give? What common experiences can you draw on to help you with your guess?

3. Knowing the density of an object can be very useful. The following situation explains why this is so.

   A geologist in the desert finds a huge rock in the shape of a *cube* (Figure 2-4). He would like to know the weight of it. However, since the rock is as big as a three-story building, he can't very well weigh it on his triple-beam balance. What can he do?

   The geologist can solve his problem by finding the density of the rock. If he knows the density, then he can simply determine the volume of the rock, using the formula for the volume of a cube ($V = s^3$, where s is the length of one side of the cube). He can determine the dimensions of the cube by using a tape measure.

$$D = \frac{m}{V} \qquad \text{therefore} \qquad m = D \times V$$

   If he doesn't know the density of the rock, the geologist can find that out by chipping a few pieces from the rock and weighing these pieces on his triple-beam balance. Next he can find the volume of his few pieces of rock by the method of water displacement. From this information, he can compute the rock's density. Having computed the density of the rock, he can now go back and compute the mass of the large piece of rock after he measures it.

   Given the following data, compute the mass of the large cube of rock. Report your answer in kilograms and pounds.

Mass of rock chips = 50.00 g

Volume of chips = 25.0 cm^3

Length of edge of rock cube = 10.0 m

Report on Experiment 2

Name _____

Section _____ Date _____

Instructor _____

PART 1: THE TRIPLE-BEAM BALANCE

Object	Weight of Object
10-mL graduated cylinder	
25-mL graduated cylinder	
50-mL beaker	
100-mL beaker	
Aluminum weighing cup	

PART 2: DETERMINING THE DENSITIES OF SOME COMMON OBJECTS

DENSITY OF WATER

Weight of graduated cylinder + water	
Weight of graduated cylinder	
Weight of water	
Volume of water	
Density of water	

Show calculation for density here: $D = \dfrac{m}{V}$

DENSITY OF WOOD

Weight of wood block	
Length of wood block	
Height of wood block	
Width of wood block	
Volume of wood block	
Density of wood block	

Show calculations for volume and density of wood block here:

$$V = l \times w \times h \qquad\qquad D = \frac{m}{V}$$

DENSITY OF IRON (BY DIRECT MEASUREMENT)

Weight of iron cylinder	
Height of iron cylinder	
Diameter of iron cylinder	
Volume of iron cylinder (calculated)	
Density of iron cylinder	
Density of iron cylinder from *Handbook of Chemistry and Physics*	

Show calculations for volume and density here:

$$V = \frac{\pi d^2 h}{4} \qquad\qquad D = \frac{m}{V}$$

Report on Experiment 2

Name _____

Section _____ Date _____

Instructor _____

DENSITY OF IRON (BY WATER DISPLACEMENT)

Weight of iron cylinder	
Initial reading of graduated cylinder	
Final reading of graduated cylinder	
Volume of iron cylinder (by water displacement)	
Density of iron cylinder	

Show calculation for density here: $D = \dfrac{m}{V}$

DENSITY OF GLASS MARBLE

Weight of glass marble	
Initial reading of graduated cylinder	
Final reading of graduated cylinder	
Volume of glass marble	
Density of glass marble	

Show calculation for density here: $D = \dfrac{m}{V}$

DENSITY OF ETHYL ALCOHOL

Weight of graduated cylinder + ethyl alcohol	
Weight of graduated cylinder	
Weight of ethyl alcohol	
Volume of ethyl alcohol	
Density of ethyl alcohol	
Density of ethyl alcohol from *Handbook of Chemistry and Physics*	

Show calculation for density here: $D = \dfrac{m}{V}$

RESPONSES TO QUESTIONS

1.

2.

3.

Experiment 3
Separation of Solids From Liquids

Time: About 2 hours

Materials: $1M$ lead nitrate solution, $0.2M$ potassium chromate solution, $0.25M$ barium chloride solution, $3M$ sulfuric acid, $0.2M$ aluminum nitrate solution, $6M$ ammonium hydroxide solution, filter paper, Büchner funnel, suction flask

Safety and Chemical Hazard Guide

Please review this before performing the experiment.

1. Glasses must be worn at *all* times when in the laboratory!
2. Lead nitrate solution. Avoid ingestion, which can cause lead poisoning. Avoid contact with skin. If contact is made, wash affected area immediately with soap and water. Wash hands after handling solution. Lead solutions affect the central nervous system, gastrointestinal tract, kidneys, and blood.
3. Potassium chromate solution. Avoid ingestion. This chemical can stain skin on contact. Wash affected area immediately.
4. Barium chloride solution. Avoid ingestion. This chemical can be a skin irritant. Wash affected area immediately.
5. Sulfuric acid. CAUTION. This acid can cause skin burns on contact. Wash affected area immediately with plenty of water and notify your instructor. It is an eye, nose, and throat irritant, so avoid inhalation. Its target organs are the respiratory system, eyes, skin, and teeth.
6. Aluminum nitrate solution. No unusual hazards. Avoid ingestion.
7. Ammonium hydroxide solution. CAUTION. This chemical can cause skin burns on contact. Wash affected area immediately with plenty of water and notify your instructor. It is an eye, nose, and throat irritant. Avoid inhalation and ingestion. Inhalation of this substance can cause bronchial spasms (coughing), chest pain, and pulmonary edema. Its target organs are the eyes, lungs, and respiratory system.

INTRODUCTION

Separating solids from liquids is a common operation in most laboratory work. The chemist uses various methods to perform this task. The method the chemist chooses depends on the nature of the solid and liquid he or she is trying to separate.

Some Things You Will Learn by Doing This Experiment

1. You will learn to separate solids from liquids by using the following techniques:
   a) Gravity filtration
   b) Suction filtration
   c) Centrifuging
   d) Decanting
2. You will be able to figure out the best separation technique once you know the nature of the solid and liquid.

DISCUSSION

In this experiment your instructor will demonstrate three chemical reactions. Each of these reactions produces a solid substance that precipitates from the solution. The three solids that will be formed are lead chromate, barium sulfate, and aluminum hydroxide.

$$Pb(NO_3)_2 + K_2CrO_4 \rightarrow 2KNO_3 + PbCrO_4$$

Lead(II) nitrate + Potassium chromate → Potassium nitrate + *Lead(II) chromate*

$$BaCl_2 + H_2SO_4 \rightarrow 2HCl + BaSO_4$$

Barium chloride + Sulfuric acid → Hydrochloric acid + *Barium sulfate*

$$Al(NO_3)_3 + 3NH_4OH \rightarrow 3NH_4NO_3 + Al(OH)_3$$

Aluminum nitrate + Ammonium hydroxide → Ammonium nitrate + *Aluminum hydroxide*

Each of these precipitates is very different. The lead chromate precipitate appears to be composed of large crystals when compared to the barium sulfate precipitate, which appears to be composed of very fine crystals. The aluminum hydroxide precipitate appears to be a gelatinous material.

In the next section of this experiment you will be given samples of each of these precipitates, along with the solutions from which they were formed. It will be your job to separate each of the precipitates from the solutions by using various methods. Each method will be explained in detail as you proceed. After completing the experiment, you will be asked to compare the effectiveness of each method in terms of separating each solid from its solution.

PART 1: PREPARING FOR THE EXPERIMENT

PROCEDURE

. . . .

Note: Put on your safety glasses.

. . . .

Step 1
Filter paper

Step 2
Filter paper is
folded in half

Step 3
Filter paper
is folded in
half again

Step 4
Pull three sides
of the filter paper
away from the
fourth side to
form the cone

Figure 3-1
How to fold filter paper

1. Obtain a test-tube rack and twelve clean, dry test tubes.
2. Label four of the test tubes lead chromate.
3. Label four of the test tubes barium sulfate.
4. Label four of the test tubes aluminum hydroxide.
5. Your instructor will either dispense four samples of each solution or give you instructions on how to prepare the precipitates yourself.

PART 2: SEPARATION BY GRAVITY FILTRATION

PROCEDURE

1. Obtain three pieces of filter paper and fold each as shown in Figure 3-1.
2. Place each piece of folded filter paper in a separate funnel, as shown in Figure 3-2. Wet each piece of filter paper with water so it adheres to the sides of the funnel.

. . . .

Note: The general practice is always to wet the filter paper with the solvent you're going to pour through it.

. . . .

Figure 3-2
Filter paper positioned in funnel

3. Into the first filtering setup, pour the contents of one test tube containing the lead chromate solution.

4. Into the second filtering setup, pour the contents of one test tube containing the barium sulfate solution.

5. Into the third filtering setup, pour the contents of one test tube containing the aluminum hydroxide solution.

6. Observe the relative rate of filtration of each solution. Also note the efficiency of separation for each of the three substances. Record your observations.

PART 3: SEPARATION BY SUCTION FILTRATION

PROCEDURE

1. Obtain a Büchner (būk-ner) funnel and a filtering flask. Assemble the apparatus as shown in Figure 3-3. Note that the suction flask is clamped to the ring stand. Be sure to use the proper rubber tubing to connect the flask to the water aspirator. This tubing usually has thick walls that won't collapse during vacuum filtration.

2. Obtain three pieces of filter paper. Make sure their diameter is such that they fit your Büchner funnel correctly. (Filter paper that fits correctly should lie flat in the Büchner funnel.)

3. Place one piece of filter paper in the Büchner funnel and wet it with water, so it stays flat on the surface of the funnel.

4. Turn on the water aspirator and filter the contents of one test tube containing the lead chromate solution. After filtering is completed, turn off the water aspirator, using the following procedure:

   a) Disconnect the tubing at the aspirator connection.

   b) Turn off the water.

5. Clean the filtering apparatus. Place a new piece of filter paper in the funnel and filter the contents of one test tube containing the barium sulfate solution.

6. Repeat the procedure of cleaning the apparatus. Filter the contents of one test tube containing the aluminum hydroxide solution.

7. Notice the rate and efficiency of separation for each of the three substances. Record your observations.

Filter paper

Büchner funnel

Rubber stopper

Connect to rubber tubing that connects to water aspirator

Filtering flask

Figure 3-3
Equipment needed for a suction filtration

PART 4: SEPARATION BY CENTRIFUGING

PROCEDURE

1. Your instructor will explain the operation of the centrifuge you will use.
2. Bring one sample of each solution in its own test tube to the centrifuge.
3. Place each test tube in the centrifuge, and arrange the test tubes so that the centrifuge will be balanced during operation.

. . . .

Note: You may have to transfer each of the solutions into special centrifuge test tubes.

. . . .

4. Allow the test tubes to centrifuge for about two minutes.

. . . .

Caution: After turning the power off, allow the centrifuge wheel to stop by itself. Do *not* place your hands or fingers near the wheel until it stops spinning.

. . . .

5. Remove the test tubes. Notice the efficiency of separation of this method as compared with the other methods you have tried so far. Record your observations.

PART 5: SEPARATION BY DECANTING

PROCEDURE

1. At this point in the experiment, you should have one test tube of each solution remaining. Because you have left these remaining test tubes standing undisturbed for the entire laboratory period, each of the solids in the test tubes has had a chance to settle. See if you can pour off the liquid portion from each test tube without causing loss of the solid. This procedure is separation by decantation.
2. Compare this method of separation with other methods used in this experiment. Record your observations.

QUESTIONS

Questions 1 through 4: For each of the following questions, choose the method you found to work best: gravity filtration, suction filtration, centrifuging, or decantation.

1. Which method separated all three types of precipitates?
2. Which method, in general, was the least efficient in terms of separation of the solid from the liquid?
3. Which method, in general, was the most rapid?
4. Which method was the least costly in terms of the equipment needed?

Questions 5 through 8: For each of the following questions, list the names of the precipitates that best fit the answer: lead chromate, barium sulfate, or aluminum hydroxide.

5. Which precipitate(s) could not be separated by gravity filtration?
6. Which precipitate(s) could not be separated by suction filtration?
7. Which precipitate(s) could not be separated by centrifuging?
8. Which precipitate(s) could not be separated by decantation?

Place your answers to Questions 1 through 8 on the report page.

Report on Experiment 3

Name _____

Section _____ Date _____

Instructor _____

OBSERVATIONS ON THE FOUR METHODS OF SEPARATING SOLIDS FROM LIQUIDS

RESPONSES TO QUESTIONS

1.

2.

3.

4.

5.

6.

7.

8.

©1988 Houghton Mifflin Company

Experiment 4
Use of the Gas Burner: Study of Elements, Compounds, and Mixtures

Time: About 2 hours

Materials: Gas burner, striker, piece of cardboard, crucible, crucible tongs, bar magnet, selection of various substances in test tubes labeled by letter code

Safety and Chemical Hazard Guide

Please review this before performing the experiment.

1. Glasses must be worn at *all* times when in the laboratory!
2. Mercury(II) oxide (substance in test tube A). CAUTION. Avoid inhalation, ingestion, and skin contact with this substance or mercury poisoning can result. Wash skin immediately if contact occurs. The target organs of this substance are the skin, respiratory system, central nervous system, and kidneys.
3. Substance in test tube B. No hazards.
4. Substance in test tube C. No hazards.
5. Cobalt(II) chloride solution (substance in test tube D). Avoid ingestion and skin contact. If ingested in large amounts, this substance can depress white blood cell production. If ingested it can also cause flushing, chest pains, skin rashes, nausea, and vomiting.
6. Iron and sulfur (substance in test tube E before heating). Iron metal has no special hazards. Sulfur can cause irritation of the skin and mucous membranes. Avoid ingestion and inhalation.
7. Iron(II) sulfide (substance in test tube E after heating). Use the same precautions as with sulfur. Also, do not mix this substance with acids in an uncontrolled environment, because it will liberate poisonous hydrogen sulfide gas.

INTRODUCTION

In Experiment 1 you were given a list of chemicals that you were asked to classify. The means you had at your disposal for classification were limited. You were not allowed to touch the chemicals or perform any chemical tests on them. You were allowed only to observe the substances and test them with a magnet. Your classification, therefore, had to be based on criteria such as:

a) Physical state of the substance (solid, liquid, gas)

b) Color of the substance

c) Texture of the substance

d) Magnetic properties of the substance

In this experiment you are going to work with some of these same substances, and you are again going to try to classify substances. The scheme you will use will be the one most chemists use today. However, in order to be able to use this classification scheme and to decide where a particular substance belongs, you will have to test each substance. Many of the tests will involve heating the material to be tested. It is for this reason that our laboratory experiment begins with a section on how to use the gas burner.

Some Things You Will Learn by Doing This Experiment

1. You will learn parts of the laboratory gas burner.

2. You will learn how to use a gas burner.

3. You will learn the difference between an element, a compound, a mixture, and a solution.

4. You will learn how to decompose certain compounds.

5. You will learn how to separate a solution.

DISCUSSION

PART 1: USE OF THE GAS BURNER

There are various types of gas burners used in chemistry laboratories today (Figure 4-1). However, almost all gas burners have similar parts.

1. Gas inlet. The gas inlet allows the gas to enter the burner from the supply valve on the laboratory bench.

2. Base. The base supports the burner so it stands erect.

3. Gas adjustment screw. The gas adjustment screw enables the operator to control the amount of gas entering the burner. Some burners do not have a gas

Figure 4-1
Two types of laboratory gas burners

adjustment screw. One can control the gas flow on these burners only by using the gas-flow supply valve on the laboratory bench.

4. *Barrel.* The barrel is the part of the burner in which the gas and air mix.

5. *Air inlet.* The air inlet enables the operator to vary the amount of air entering the barrel. (It's like the carburetor in a car.) There are various types of air inlets on different types of gas burners. On some burners, the control is located on the base of the burner, which pushes in or pulls out. On other burners a collar that can turn is located around the base of the barrel. Still others have barrels that contain holes near the base. In these types of burners, the whole barrel turns. All three types of air inlets work on the same principle. They allow the operator to increase or decrease the passageway through which the air enters the barrel.

PROCEDURE

. . . .

Note: Put your safety glasses on!

. . . .

1. Obtain your gas burner.
2. Identify the parts of the burner and familiarize yourself with them.
3. Connect the burner to the supply outlet on the laboratory bench with rubber tubing.
4. Open the air inlet on the burner about halfway.
5. If your burner has a gas inlet, open it halfway.
6. With your safety glasses on and your hair out of the way of the burner, turn the gas-supply valve on fully.
7. Using a striker (or match), light the burner near the top of the barrel (Figure 4-2).
8. Look at the flame. It should be 4 to 6 cm high. If it isn't, adjust the air flow.

Figure 4-2
Lighting a gas burner with a striker

. . . .

Caution: If the gas starts burning inside the barrel, turn off the gas supply immediately. This is known as *flashback. Do not* touch the barrel; it will probably be very hot. Allow the burner to cool. Before relighting, adjust the gas-and-air mixture by doing one of the following things:

   a) If your burner has a gas-inlet control, increase the gas supply.

   b) Decrease the air supply.

. . . .

9. If the flame goes out after you have lighted it, turn off the gas supply and increase the air supply before relighting.

10. Observe the different types of flames you can obtain with the burner by first slowly decreasing the air supply and then slowly increasing the air supply.

11. When you allow only a little air to enter the barrel of the burner, you get a yellow flame. This means that the gas is not being combusted (burned) completely, and carbon soot is being produced.

12. To demonstrate that soot is being produced, obtain a porcelain crucible and hold it with a crucible tongs in the yellow flame of the burner. A black deposit of soot should appear on the crucible.

13. When the air supply is fully opened, you get a blue flame.

14. At optimum conditions you can obtain a blue, almost invisible flame, which means that the gas is being burned most completely. The flame at this point should appear like the one shown in Figure 4-3. This is the flame you will use in most of your laboratory work.

15. With the flame set at optimum conditions, observe the three cones that compose the flame.

   a) The cone closest to the barrel. This cone contains the mixture of gases before it starts to burn.

   b) The middle cone. In the middle cone the gases are burning, but not completely.

   c) The outer cone. The gases are burning completely in the outer cone, and it is therefore the hottest part of the flame.

16. Obtain a wood splint and place it in the cone closest to the barrel. Does the splint burn?

17. Obtain a piece of cardboard and place it in the flame, so that one edge of the cardboard rests on the barrel of the burner. Allow the cardboard to be in the

Gas burner flame
at optimum conditions

Figure 4-3
Gas burner flame at optimum conditions

Figure 4-4
Classification of matter

flame only for an instant, just long enough to obtain a scorch pattern. Describe the results in the data table on your report sheet, located at the end of this experiment. After you complete this procedure, be sure to saturate the cardboard with water to eliminate the possibility of a fire in the waste receptacle.

PART 2: THE STUDY OF ELEMENTS, COMPOUNDS, AND MIXTURES

Although there are three physical states of matter—solid, liquid, and gas—matter is usually classified into two major categories, heterogeneous (having parts with different properties) and homogeneous (having similar properties throughout). These two categories can be further broken down (Figure 4-4). Below is a brief review of the major classifications.*

a) *Elements.*   These are the basic building blocks of matter. They cannot be broken down into simpler substances by ordinary chemical means.

b) *Compounds.*   These are substances that are composed of two or more elements, chemically combined in definite proportions. They can be broken down into their respective elements by chemical means.

c) *Heterogeneous mixtures.*   Heterogeneous mixtures consist of two or more substances, each of which retains its own characteristic properties. Mixtures can be separated by physical means.

d) *Solutions.*   These are homogeneous mixtures whose compositions can be varied within certain limits. Solutions are mixtures of compounds.

In this experiment you will be given a series of substances. You will try to classify each substance as one of these four types of matter.

PROCEDURE

a) SUBSTANCE A**

1. Obtain a set of substances to be classified.
2. Observe the material in test tube A and note its color. Does the material seem to have the same consistency throughout?
3. Let's see whether this material is an element or a compound.

*You can find a complete discussion of the different types of matter in Chapter 3 of our text, *Basic Concepts of Chemistry* (Boston: Houghton Mifflin, 1988).

**Your instructor may want to perform this part of the experiment as a demonstration.

. . . .

Note: It is very hard to prove that a given substance is an element. For this experiment we shall use an indirect type of proof. We shall start out by assuming that a substance is an element. We'll then try to break the substance down into simpler substances. If we are able to do this, we may then assume that the original substance wasn't an element.

. . . .

4. Place test tube A in a test tube holder, take it to a fume hood, and heat it with your gas burner or hot plate.

5. Note that the substance in the test tube changes color. More important is the appearance of a shiny metallic type of substance on the upper walls of the test tube.

6. The substance you are working with is called mercury(II) oxide. What do you think is the substance coating the sides of the test tube?

7. If the mercury(II) oxide is decomposing, you could test for the other material being released—oxygen gas.

8. To test for oxygen gas, obtain a wood splint, ignite it, and then put out the flame, so that the splint glows. While you continue to heat the mercury(II) oxide, place the glowing splint into the test tube. If oxygen gas is being given off, the glowing splint should reignite.

9. Classify the mercury(II) oxide in the data table.

b) SUBSTANCE B

1. Observe the material in test tube B.

2. Does it appear to have a similar consistency throughout, or does it appear to be composed of two or more substances?

3. Add enough water to test tube B to fill it halfway. Shake the contents of the test tube vigorously. What happens?

4. Filter the contents of test tube B (Figure 4-5) and collect the liquid portion (the filtrate) in an evaporating dish as it passes through the filter paper. (Recall the filtering you did in Experiment 3.)

5. Examine the material left on the filter paper. Observe its color and texture. Can you determine what it is?

6. To recover the material that dissolved in the water, boil off the water collected in the evaporating dish (Figure 4-6).

7. After the evaporating dish has cooled, examine the material in the dish. Is it different from the material on the filter paper?

8. Classify the contents of test tube B.

Step 1
Filter paper

Step 2
Filter paper is
folded in half

Step 3
Filter paper
is folded in
half again

Step 4
Pull three sides
of the filter paper
away from the
fourth side to
form the cone

Figure 4-5
Setting up equipment to filter contents of test tube B

Figure 4-5
(continued)

c) SUBSTANCE C

1. Observe the contents of test tube C. Note that it is a piece of wire attached to a piece of glass. It is the *wire* that we are interested in classifying.
2. Remove the glass tube and wire from the test tube. Using the glass tube as a handle, hold the wire in the flame of the burner for about one minute.

. . . .

Note: *Do not touch the hot wire!*

. . . .

3. Allow the wire to cool and observe it. Do you see any change in it? The wire is made out of platinum.
4. Classify the wire in test tube C.

c) SUBSTANCE D

1. Observe and describe the contents of test tube D.

Figure 4-6
Boiling off water in an evaporating dish

2. Pour the contents of test tube D into an evaporating dish and boil off the liquid. Describe what remains.

3. Classify the contents of test tube D.

e) SUBSTANCE E

1. Observe the contents of test tube E. Does the material appear to be of similar consistency throughout, or is it composed of two or more substances?

2. Obtain a bar magnet and place one end of it against the side of the test tube. Move the bar magnet slowly up the side of the test tube. What happens?

3. Classify the contents of test tube E.

4. Remix the two materials in the test tube and take the test tube to the fume hood.

5. Heat the contents of the test tube in the fume hood until you observe a bright glow. Remove the test tube from the flame and allow its contents to cool.

6. Observe the contents of the test tube. Ask your instructor to help you remove the contents from the test tube.

. . . .

Note: Your instructor may have you do this yourself.

. . . .

The original materials in the test tube were iron and sulfur. Can you separate the iron and sulfur with a magnet at this point, as you did in Step 2?

7. Classify the substance you have formed in test tube E after heating.

QUESTIONS

1. If a match were placed in the cone closest to the barrel of the gas burner, do you think that it would ignite? Explain.

2. How would you go about separating a solution composed of two liquids, for example, ethyl alcohol and water? (*Hint:* See the *Handbook of Chemistry and Physics,* which is published annually by the Chemical Rubber Company.)

3. Many compounds do not decompose into their elements when they are heated. For example, water simply changes to steam when heated. What are some other methods you might use to break water into its elements?

4. In this experiment we heated a piece of platinum metal and noted that it did not change chemically. We therefore assumed that it was an element. However, this type of indirect proof can lead to a disaster if it is applied generally, because of the following facts:

   a) Some elements change their physical state when they are heated. This process could be misinterpreted as a compound breaking down.

   b) When you heat some elements, they combine with oxygen in the air. For example, magnesium forms magnesium oxide when heated. This process could also be misinterpreted as a compound breaking down.

   Suppose that you are given an unknown solid substance. Describe some tests you might perform to prove that it is either an element or a compound.

Report on Experiment 4

Name _____

Section _____ Date _____

Instructor _____

PART 1: USE OF THE GAS BURNER

Describe the results of heating the crucible in the yellow flame.

Describe the results of placing the wood splint in the cone closest to the barrel of the burner.

Describe the results obtained by placing the cardboard in the flame, and show a scorch-pattern diagram.

PART 2: THE STUDY OF ELEMENTS, COMPOUNDS, AND MIXTURES

a) SUBSTANCE A

1. Describe the color, physical state, and so forth, of the substance.

2. Describe what happens to the substance when it is heated.

3. What is the shiny metallic substance coating the sides of the test tube?

4. Describe the results of the oxygen test.

5. Classify substance A as an element, compound, mixture, or solution.

b) SUBSTANCE B

1. Describe the color, physical state, and so forth of the substance.

2. Describe what happens when you add water to the contents of the test tube.

3. Describe the appearance of the material left on the filter paper (color, texture, and so forth). Can you determine what it is?

4. Describe the substance recovered from the evaporating dish (color, texture, and so forth).

5. Classify substance B as an element, compound, mixture, or solution.

Report on Experiment 4

Name _____

Section _____ Date _____

Instructor _____

c) SUBSTANCE C

1. Describe the color, physical state, and so forth, of the substance.

2. Describe the substance after it has been heated.

3. Classify substance C as an element, compound, mixture, or solution.

d) SUBSTANCE D

1. Describe the color, physical state, and so forth, of the substance.

2. Describe what remains in the evaporating dish after you boil off the liquid.

3. Classify substance D as an element, compound, mixture, or solution.

e) SUBSTANCE E

1. Describe the color, physical state, and so forth, of the substance.

2. What happens when you apply a magnet to substance E?

3. Classify substance E as an element, compound, mixture, or solution.

4. Describe the material in test tube E after it has been heated.

5. Classify the new substance in test tube E that has been formed by the heating.

Summary of Results

Substance in Test Tube	Classification
A	
B	
C	
D	
E	
E (after heating)	

Report on Experiment 4

Name _____

Section _____ Date _____

Instructor _____

RESPONSES TO QUESTIONS

1.

2.

3.

4.

Experiment 5
Empirical Formula of a Compound

Time: About 2 hours

Materials: Magnesium ribbon, crucible plus cover, crucible tongs, gas burner, balance

Safety and Chemical Hazard Guide

Please review this before performing the experiment.

1. Glasses must be worn at *all* times when in the laboratory!
2. Magnesium ribbon. There are no special hazards, but inhalation of magnesium dust can be irritating to the respiratory system.
3. Magnesium oxide (formed after heating magnesium). This chemical may be irritating to the eyes, lungs, and respiratory system. Inhalation of the fumes or ingestion of the dust can cause flulike symptoms, including fever, cough, and chest pains.

INTRODUCTION

In research laboratories all over the world, it is the job of the chemist to figure out the chemical formulas of unknown compounds. The chemist's first task is to find out what elements compose the compound. The next step is to find the ratio of the atoms of each element in a molecule of the compound. This enables the chemist to determine the empirical formula of the compound.

> *The empirical formula is the simplest whole-number ratio of atoms in a molecule of the compound. Once the chemist knows the empirical formula of a compound, he or she can try to discover its molecular formula. The molecular formula of a compound is the number of atoms of each element in a molecule of the compound.*

Some Things You Will Learn by Doing This Experiment

1. You will learn what an empirical formula of a compound is, and what a molecular formula of a compound is.

2. You will learn how to obtain data experimentally, and how to use these data to determine the empirical formula of magnesium oxide.

DISCUSSION

Suppose that you are a chemist, and someone gives you 10.00 g of an unknown compound, asking you to find out what it is. What do you do?

You perform an elemental analysis of the compound and discover that it is composed of carbon, hydrogen, and oxygen. Upon further analysis, you find that from your 10.00 g of compound you can obtain 4.00 g of carbon, 0.67 g of hydrogen, and 5.33 g of oxygen. Now look up the atomic weight of each element in the periodic table. You will find that carbon's atomic weight is 12, hydrogen's is 1.0, and oxygen's is 16. Using all this information, you find the empirical formula of the compound as follows:

$$\text{Moles of C atoms} = (4.00 \text{ g})\left(\frac{1.0 \text{ mole}}{12 \text{ g}}\right) = 0.33 \text{ mole}$$

$$\text{Moles of H atoms} = (0.67 \text{ g})\left(\frac{1.0 \text{ mole}}{1.0 \text{ g}}\right) = 0.67 \text{ mole}$$

$$\text{Moles of O atoms} = (5.33 \text{ g})\left(\frac{1.0 \text{ mole}}{16 \text{ g}}\right) = 0.33 \text{ mole}$$

$C_{0.33}H_{0.67}O_{0.33}$ or in terms of simplest *whole* numbers,

$C_{1.0}H_{2.0}O_{1.0}$ or just CH_2O (obtained by dividing by the smallest number
in $C_{0.33}H_{0.67}O_{0.33}$)

The empirical formula of the compound is CH_2O.

If you wish to find the molecular formula of the compound, you must obtain its molecular weight. Let's say that the molecular weight of this compound has been determined and found to be 180. How can we use this information, along with the information about the empirical formula, to obtain the molecular formula of the unknown compound? First we determine the formula weight of CH_2O.

. . . .

Note: The atomic weight of C is 12, H is 1, and O is 16.

. . . .

Formula weight of $CH_2O = (1)(12) + (2)(1) + (1)(16) = 30$.

The molecular formula is a whole-number multiple of the empirical formula.

Determine what it is for this compound.

$30x = 180$; therefore $x = 6$.

What did we do here? We used the following logic.

1. The empirical formula is CH_2O.
2. The molecular formula must have this same carbon-hydrogen-oxygen ratio; in other words, 1 to 2 to 1, but it must have a molecular weight of 180. This means $(CH_2O)_x = 180$ or $30x = 180$, or $x = 6$.
3. Since the formula weight of CH_2O is 30, the molecular formula must have six times as many carbons, hydrogens, and oxygens per molecule as the CH_2O unit. Therefore the molecular formula of the compound is $(CH_2O)_6$ or $C_6H_{12}O_6$.

In this experiment, you are going to react a piece of magnesium ribbon with oxygen to form a compound. You will do this by heating the magnesium ribbon in air. Then you will determine the empirical formula of the resulting compound, magnesium oxide, using a method analogous to the one we have just described.

PROCEDURE

1. Obtain a crucible and cover. Be sure that it is clean and dry. (If it is not, your instructor will explain a procedure to you called "heating to constant weight.")
2. Weigh the crucible plus cover and record the weight to the nearest 0.01 g in the data table on the report sheet at the end of this experiment.
3. Get a piece of magnesium ribbon from your instructor. Clean it with a dry paper towel to remove moisture and oil that may have gotten on it from handling. (Your instructor may also have you clean the magnesium ribbon with sandpaper to remove any oxide coating the surface of the ribbon.)
4. Roll the magnesium ribbon into a loose coil. Do this neatly and carefully. The size of the coil should be such that it will fit the bottom of the crucible.
5. Place the magnesium coil in the crucible and weigh the crucible plus cover plus magnesium. Record this weight in the data table.
6. Compute the weight of the magnesium by subtraction.
7. Obtain a ring stand, an iron ring, a clay triangle, and a gas burner.
8. Set up the equipment as shown in Figure 5-1.
9. Heat the crucible and its contents, using the hottest part of the burner flame. Be certain that the cover is on the crucible. Heat for 15 minutes.
10. After you have heated the crucible for 15 minutes, obtain a crucible tongs and adjust the crucible cover so that a slight opening exists to let more air enter the crucible (Figure 5-2). Continue heating for an additional 15 minutes.
11. Shut off the gas burner and allow the crucible to cool for 10 minutes. During this initial heating, magnesium metal combined with oxygen in the air to form magnesium oxide. However, also during this initial heating, magnesium metal combined with nitrogen in the air to form magnesium nitride. In other words, the crucible at this point contains a mixture of magnesium oxide and magnesium nitride. This could cause a serious error in the determination, unless the magnesium nitride is converted to magnesium oxide. The procedure which follows solves this problem.

© 1988 Houghton Mifflin Company

— Crucible plus cover

— Clay triangle

— Iron ring

Figure 5-1
Setup for determination of empirical formula

Figure 5-2
Adjusting the crucible cover

12. Using an eyedropper, add about 40 drops (2 mL) of water to the contents of the crucible.

13. Heat the crucible again with the cover fully on for an additional five minutes. Heat gently at first.

14. Turn off the gas burner. Allow the crucible to cool for 10 minutes and weigh it. Record the weight of the crucible plus cover plus magnesium oxide in the data table.

. . . .

Note: The crucible should be cool enough to hold in your hand before you weigh.

. . . .

15. During steps 12 and 13, the magnesium nitride converted to magnesium oxide in the following manner:

    a) Magnesium nitride + water → magnesium hydroxide + ammonia gas

    b) Magnesium hydroxide + heat → magnesium oxide + water vapor

16. From the data obtained, calculate the empirical formula of magnesium oxide.

17. Obtain a second set of data from your partner and record it in the data table. Compare your results with your partner's results.

QUESTIONS

1. If you had used a different weight of magnesium metal, would you expect the empirical formula of the compound to be the same? Explain.

2. If you had omitted steps 12 and 13 from the experiment, how would this omission have affected your results in terms of the empirical formula of the magnesium oxide? In other words, would the ratio of the magnesium to the oxygen have been greater or less than the obtained ratio?

3. From the following raw data, calculate the empirical formula of the compound.
    A chemist heats 50.00 g of sulfur under controlled conditions to produce a sulfur-oxygen compound. The weight of the sulfur-oxygen compound is 100.00 g. What is the empirical formula of the sulfur-oxygen compound?

Report on Experiment 5

Name _____

Section _____ Date _____

Instructor _____

Quantity	How to Obtain	Your Values	Partner's Values
1. Weight of crucible + cover	Weigh		
2. Weight of crucible + cover + magnesium	Weigh		
3. Weight of magnesium	(2) − (1)		
4. Weight of crucible + cover + magnesium oxide	Weigh		
5. Weight of magnesium oxide	(4) − (1)		
6. Weight of oxygen	(4) − (2)		
7. Atomic weight of magnesium	Periodic Table	24.3	24.3
8. Atomic weight of oxygen	Periodic Table	16.0	16.0
9. Moles of magnesium atoms	(3) ÷ (7)		
10. Moles of oxygen atoms	(6) ÷ (8)		
11. Empirical formula of magnesium oxide	Find simplest whole-number ratio between (9) and (10).		

CALCULATIONS

RESPONSES TO QUESTIONS

1.

2.

3.

Experiment 6
Analyzing Unseen Things (Electrons in Atoms and Objects in "Black Boxes")

Time: About 2 hours

Materials: Assorted boxes (with various compartmental arrangements in each box and different objects in each box), a magnet, nichrome wire attached to a glass rod, $6M$ HCl, sodium chloride (NaCl), calcium chloride (CaCl$_2$), potassium chloride (KCl), barium chloride (BaCl$_2$), strontium chloride (SrCl$_2$), lithium chloride (LiCl), a gas burner, a cobalt-blue glass plate

Safety and Chemical Hazard Guide

Please review this before performing the experiment.

1. Glasses must be worn at *all* times when in the laboratory!
2. Hydrochloric acid ($6M$). CAUTION. This acid can cause severe burns on skin, as well as in eyes, nose, and throat. Coughing, choking, and skin rashes can result. Avoid inhalation, ingestion, and skin contact. Wash affected area immediately with plenty of water and notify your instructor.
3. Sodium chloride. No special hazards.
4. Calcium chloride. No special hazards. However, avoid ingestion, inhalation, and skin contact.
5. Potassium chloride. No special hazards. However, large doses by mouth can cause gastrointestinal irritation, vomiting, weakness, and circulatory disturbances.
6. Barium chloride. CAUTION—poisonous substance! Avoid ingestion, inhalation, and skin contact. This substance causes upper respiratory irritation, gastrointestinal upset, muscle spasms, and slow pulse. It can also cause irritation and burns to the skin and eyes.
7. Strontium chloride. No special hazards. However, avoid ingestion, inhalation, and skin contact.
8. Lithium chloride. Avoid ingestion, inhalation, and skin contact. Prolonged absorption may cause disturbed electrolyte balance, impaired renal function, and central nervous system disturbances.

INTRODUCTION

One of the major concepts you meet in chemistry is that of the atom. The concept of the atom underlies the main principles of chemical science. However, atoms are so small that scientists cannot observe them directly. How do we know they exist? Scientists rely on indirect evidence. They perform numerous experiments, and from the results they try to develop a picture of what the atom should look like and how it should behave.

Some Things You Will Learn by Doing This Experiment

1. You will learn how a scientist uses the scientific method to design a model for substances that cannot be seen.
2. You will learn how to perform flame tests.
3. You will learn how to use a scientific model to gain further information about atoms. For example, the movement of electrons in energy levels permits the identification of certain ions whose presence is shown by flame tests.

DISCUSSION

The scientist first develops a model, then determines if it will stand the test of time. Does this model account for present-day observed properties of the atom? Will it account for all future properties that might be observed? It is this process that has refined our modern concept of the atom.

In this experiment you will have the chance to try to describe some things which you cannot see directly. By doing this, you will be able to discover for yourself how a scientific model is developed.

The experiment is divided into two parts. In the first part, you will be given a box which contains an object or objects. You will try to describe the object(s) as well as the interior of the box. In the second part, you will be given some chemical compounds. You will use flame tests to identify some of the ions that compose these compounds. You will then be given an unknown compound containing one of the ions you just tested for, and you will try to identify the unknown ion by performing the flame test.

PART 1: THE "BLACK BOX" EXPERIMENT

PROCEDURE

Your instructor will give you a sealed box. This box contains an object or a group of objects. The interior of the box may or may not be partitioned. It will be your job to answer the following questions.

1. Describe the internal arrangement in the box. Are there partitions? If so, how many, and how are they arranged?
2. How many objects are in the box? Are all the objects in the same compartment?
3. What is the shape of the objects?
4. What is the mass, composition, color, and size of the objects?
5. Are the objects attracted to a magnet?

Perform any tests you feel are necessary to answer the preceding questions. However, you may *not* open the box! Be sure to record your observations and conclusions on the report page at the end of this experiment. Remember that you are attempting to propose a model of the interior of the box, using your observations to reinforce or destroy the proposed model. After you have completed your analysis of one box, try another.

PART 2: FLAME TESTS

When one heats the ions of certain elements in a flame, a characteristic color is obtained. The color is due to the energy released when electrons move from higher to lower energy levels in the ions. Here's what happens.

1. The electrons in atoms are arranged in various energy levels.
2. When they are heated, the electrons gain energy and move to higher energy levels.
3. After the electrons reach these higher energy levels, they fall back to their original energy levels. When the electrons fall back, they emit, or give off, energy.
4. Some of the energy emitted is in the form of visible light.

Although we are not able to see the electrons in an atom actually hopping from one energy level to another, we can observe the results of their hopping in terms of the light or energy they emit. (This is analogous to the "black box" experiment, in which we might listen to how something in the box sounds to give us a clue to its structure.)

Certain ions, when heated, emit specific colors of light that can be seen by the unaided eye. For example, the sodium ion produces bright yellow. We can use this phenomenon to identify certain ions in various compounds.

In this experiment you will observe the colors produced by some representative *metal* ions. You will then be given an unknown compound. This unknown compound will contain one of the *metal* ions you have previously tested. You will try to identify the ion present in your unknown compound.

PROCEDURE

1. Obtain a nichrome wire attached to a glass rod.
2. Obtain a small amount (about the size of a pea) of each of the following substances.
   a) Sodium chloride (NaCl)
   b) Calcium chloride ($CaCl_2$)
   c) Potassium chloride (KCl)
   d) Strontium chloride ($SrCl_2$)
   e) Barium chloride ($BaCl_2$)
   f) Lithium chloride (LiCl)
3. In a small beaker put about 20 mL of dilute, $6M$ HCl.
4. Obtain your gas burner, ignite it, and adjust the flame so that it is the hottest flame attainable.
5. Dip the nichrome wire into the HCl and then place it in the hottest part of the flame.

· ·

Note: Consult Experiment 4, Figure 4-3, to recall where the hottest part of the flame is. Be careful not to heat the glass holder.

· ·

6. Dip the wire into the HCl and then into the NaCl sample. Some of the NaCl will adhere to the wire.

7. Hold the wire in the flame and observe the color produced by the sodium ion. Record your observation in the table given on the report page at the end of this experiment.

8. Clean the wire as described in Step 5 and repeat the procedure with the other samples; record each result.

· ·

Note: When you perform the test on the potassium chloride sample, you must look at the color produced through a cobalt-blue glass. This is because most potassium compounds have sodium impurities, which mask the color produced by the potassium. The cobalt-blue glass filters out the light produced by any sodium impurity.

· ·

9. Obtain an unknown sample from your instructor and determine the metal ion which it contains by using the same procedure.

QUESTIONS

1. Today's world of science contains many "black boxes." Can you name some of them?

2. Explain the reasons for the different colors you got when you heated the compounds you examined in Part 2 of this experiment.

Report on Experiment 6

Name _____

Section _____ Date _____

Instructor _____

PART 1: THE "BLACK BOX" EXPERIMENT

Box number _____

OBSERVATIONS AND PROPOSED MODEL FOR THE INTERIOR OF THE BOX

PART 2: FLAME TESTS

Metal Ion Tested	Color Produced
Na^{1+}	
Ca^{2+}	
K^{1+}	
Sr^{2+}	
Ba^{2+}	
Li^{1+}	

Unknown number _____

Metal ion present in unknown: _____

RESPONSES TO QUESTIONS

1.

2.

Experiment 7
Odor Sensitivity

Time: About one hour to perform the experiment and one hour to discuss the results

Materials: A set of 4-ounce bottles containing samples to be tested

Safety and Chemical Hazard Guide

Please review this before performing the experiment.

Odor is an important characteristic of almost every substance. However, smelling an unknown substance can be hazardous and even fatal. Certain elements and compounds are highly toxic when they are inhaled. Anyone who plans to sniff a substance should determine well in advance whether the substance is toxic.

The method discussed in this experiment on odor sensitivity is *only* to be used for substances that have no harmful effects when a person inhales them. For this reason, we have deliberately chosen food odors.

We say all this because the standard rules for safety in chemical laboratories state that one should *never* inhale unknown chemical substances. Only in certain cases in which the odor of a *known* substance is to be determined should any chemical substance be smelled, and then one should inhale very carefully. Do this by holding the bottle containing the substance about a *foot away from your nose* and wafting the odor toward you with your hand.

Again we remind you that the procedure you will be following in this experiment *is not* a general procedure for smelling substances. It is only to be followed in determinations of odor sensitivity when dealing with substances that are safe to inhale. In general, *do not inhale chemicals in the laboratory* unless your instructor tells you to do so, and then do it only by the method we describe above.

1. Glasses must be worn at *all* times when in the laboratory!
2. Vanillin solutions (vanilla odor) (0.1%, 0.01%, and 0.001%). No special hazards with these concentrations.
3. Isopentyl acetate (banana oil) (0.1%, 0.01%, and 0.001%). No special hazards with these concentrations.

©1988 Houghton Mifflin Company

4. Butyric acid (rotten cheese odor) (0.1%, 0.01%, and 0.001%). No special hazards with these concentrations.

INTRODUCTION

Odor is in the nose of the beholder. However, it usually results from some volatile organic molecule. The source of odor can be either natural or artificial. Chemists throughout the world are involved in work pertaining to the study of odors. Their work covers areas such as:

1. The synthesis of compounds that have specific odors for the purpose of producing artificial fragrances and perfumes.
2. The synthesis of compounds capable of neutralizing the odors of other compounds.
3. Analytical methods that can quantify the concentrations or strengths of odors.

Some Things You Will Learn by Doing This Experiment

1. You will learn some of the methods and parameters used for testing and quantifying odors.
2. You will learn how to test substances for odor intensity by the ASTM triangle method.

DISCUSSION

Chemists have devised some methods and some parameters to quantify odor. One such parameter is the *threshold odor number* (TON).

The TON is the number of times a sample has to be diluted with odor-free water for the odor to be just detectable.

To determine the "just-detectable" concentration requires much training. The method used to determine the TON is called the ASTM triangle method. (The letters ASTM stand for the American Society for Testing and Materials.) This method requires the chemist to evaluate a set of three flasks, one of which contains an odorant. Chemists evaluating odorants by this method must observe three precautions.

1. Evaluators should not examine the set of flasks with the highest concentrations first because of probable odor fatigue. (This means that your nose becomes insensitive to the odor.)
2. Evaluators must work in a nearly odor-free atmosphere.
3. Evaluators must use odor-free soap to wash their hands.

In this experiment you are not actually going to determine the TON. Instead you are going to evaluate three types of odors. These odors will smell like cheese (more like a rotted cheese), banana, and vanilla. Each will occur in the test containers in various concentrations. You will have to decide which of the three containers has the odorous substance. You will also evaluate the intensity of the odorous substance on a scale of zero to five.

. . . .

Note: During the course of the experiment you must *not* compare your results with those of the other groups. Comparing will take place during a discussion session, which will be held at the end of the experiment.

. . . .

PROCEDURE

1. You will test ten triangles. (Each triangle contains three containers, but only one container in each group holds an odorous substance.)
2. Obtain the first triangle. One container holds the odorous substance, and the other two hold distilled water.
3. Take short, sharp sniffs of the sample to ensure that the odorant reaches the olfactory receptor surfaces that are high up in the nasal passages.

. . . .

Note: Remember that this smelling technique is to be used only in this experiment. It is not the general method used for detecting the odors in the chemistry laboratory.

. . . .

4. Select the container—A, B, or C—that appears to have the odor and note this on the form located on the report page.
5. Record the kind of odor (vanilla, banana, or cheese) and also the odor intensity. Odor intensity is measured on a scale from 0 to 5 with 0 indicating no odor; 1, a very faint odor; 2, a faint odor; 3, an easily noticeable odor; 4, a strong odor; and 5, a very strong odor.

Sample Test Sheet

Bottle Set	Bottle		
	A	B	C
1	—	—	Vanilla = 3

6. Wait a few minutes between examining each triangle to allow any possible effects of odor fatigue to diminish.

QUESTIONS

1. What was your major problem in evaluating the odors?
2. In addition to the TON, what other critiera can you devise to measure the strength of odorous substances? Give a working definition of your criteria.

Report on Experiment 7

Name _____

Section _____ Date _____

Instructor _____

ODOR SENSITIVITY TEST FORM

Bottle Set	Bottle Designation		
	A	B	C
1			
2			
3			
4			
5			
6			
7			
8			
9			
10			

1.

2.

Laboratory Exercise 1
Arranging Twenty-One Elements

Time: About one hour to perform, plus one hour for discussion.

Materials: 21 cards containing information about 21 elements

INTRODUCTION

For many years chemists have tried different arrangements of the elements in an attempt to find some logical order among them. In this exercise you'll try to find your *own* arrangement of the elements. You will make up a small deck of cards by cutting out the cards printed on the next page of this manual. There are 21 cards, representing 21 elements. Each card carries certain information, and your job is to find some logical way of arranging these 21 cards. Each card indicates the atomic weight, the ionization potential, and the formula for the hydride, oxide, and fluoride of the element it depicts. In your attempt to find order, look for trends among these properties. There is no "correct" way to arrange the cards, but some arrangements are better than others. After you have finished your initial arrangement, you may want to make changes until you find the best possible order. Set the cards on the table or lab area and line them up any way you choose. If you wish, use the questions that follow as guides for your arrangement.

As you become more at home in the world of chemistry, you will learn the different ways in which chemists themselves have arranged the elements. When you compare your arrangements with theirs, you will be able to note similarities and differences.

QUESTIONS

1. Is there any pattern to be found in the atomic weights?
2. Do the ionization potentials form a pattern?
3. Does a pattern exist in the hydride formulas?
4. Is there any pattern to be found in the oxide formulas?
5. Do the fluoride formulas form a pattern?

Atomic Weight: 1 No. of H in hydride: 1 No. of F in fluoride: 1 No. of O in oxide: 0.5 Ionization pot.: 314	*Atomic Weight:* 4 Hydride unknown Fluoride unknown Oxide unknown Ionization pot.: 567	*Atomic Weight:* 7 No. of H in hydride: 1 No. of F in fluoride: 1 No. of O in oxide: 0.5 Ionization pot.: 124
Atomic Weight: 9 No. of H in hydride: 2 No. of F in fluoride: 2 No. of O in oxide: 1 Ionization pot.: 215	*Atomic Weight:* 10 No. of H in hydride: 3 No. of F in fluoride: 3 No. of O in oxide: 1.5 Ionization pot.: 190	*Atomic Weight:* 12 No. of H in hydride: 4 No. of F in fluoride: 4 No. of O in oxide: 2 Ionization pot.: 260
Atomic Weight: 14 No. of H in hydride: 3 No. of F in fluoride: 3 No. of O in oxide: 2.5 Ionization pot.: 335	*Atomic Weight:* 16 No. of H in hydride: 2 No. of F in fluoride: 2 No. of O in oxide: 1 Ionization pot.: 314	*Atomic Weight:* 19 No. of H in hydride: 1 No. of F in fluoride: 1 No. of O in oxide: 0.5 Ionization pot.: 402
Atomic Weight: 20 Hydride unknown Fluoride unknown Oxide unknown Ionization pot.: 497	*Atomic Weight:* 23 No. of H in hydride: 1 No. of F in fluoride: 1 No. of O in oxide: 0.5 Ionization pot.: 119	*Atomic Weight:* 24 No. of H in hydride: 2 No. of F in fluoride: 2 No. of O in oxide: 1 Ionization pot.: 176
Atomic Weight: 45 Hydride unknown No. of F in fluoride: 3 No. of O in oxide: 1.5 Ionization pot.: 151	*Atomic Weight:* 40 No. of H in hydride: 2 No. of F in fluoride: 2 No. of O in oxide: 1 Ionization pot.: 141	*Atomic Weight:* 39 No. of H in hydride: 1 No. of F in fluoride: 1 No. of O in oxide: 0.5 Ionization pot.: 100
Atomic Weight: 40 Hydride unknown Fluoride unknown Oxide unknown Ionization pot.: 363	*Atomic Weight:* 36 No. of H in hydride: 1 No. of F in fluoride: 1 No. of O in oxide: 0.5 Ionization pot.: 300	*Atomic Weight:* 32 No. of H in hydride: 2 No. of F in fluoride: 2 No. of O in oxide: 3 Ionization pot.: 239
Atomic Weight: 31 No. of H in hydride: 3 No. of F in fluoride: 3 No. of O in oxide: 2.5 Ionization pot.: 254	*Atomic Weight:* 28 No. of H in hydride: 4 No. of F in fluoride: 4 No. of O in oxide: 2 Ionization pot.: 188	*Atomic Weight:* 27 No. of H in hydride: 3 No. of F in fluoride: 3 No. of O in oxide: 1.5 Ionization pot.: 138

Experiment 8
The Periodic Table: The Chemistry of Elements Within a Group

Time: About 2 hours

Materials: Lithium chloride, sodium chloride, potassium chloride, carbon chips, silicon chips, germanium crystals, tin strips, lead strips, strontium oxide (SrO), magnesium oxide (MgO), copper turnings, $6M$ nitric acid (HNO_3), phosphorus pentoxide (P_2O_5), arsenic pentoxide (As_2O_5), a bar magnet, neutral litmus paper, calcium oxide

Safety and Chemical Hazard Guide

Please review this before performing the experiment.

1. Glasses must be worn at *all* times when in the laboratory.
2. Lithium chloride. Avoid ingestion, inhalation, and skin contact. Prolonged absorption may cause disturbed electrolyte balance, impaired renal function, and central nervous system disturbances.
3. Sodium chloride. No special hazards.
4. Potassium chloride. No special hazards. However, large doses by mouth can cause gastrointestinal irritation, vomiting, weakness, and circulatory disturbances.
5. Carbon chips. No special hazards.
6. Silicon chips. No special hazards.
7. Germanium crystals. No special hazards.
8. Tin strips. No special hazards.
9. Lead strips. Avoid touching the strips. Wash hands thoroughly with soap and water if you touch the lead. Lead poisoning in adults is usually occupational, due mainly to inhalation of lead dust or fumes.
10. Strontium oxide. Avoid inhalation, ingestion, or skin contact. May be irritating to the eyes, nose, throat, and skin.
11. Magnesium oxide. May be irritating to the eyes, lungs, and respiratory system. Inhalation of the fumes or ingestion of the dust can cause flulike symptoms, including fever, cough, and chest pains.
12. Copper metal. No special hazards.

13. Nitric acid (6*M*). CAUTION. This acid can cause severe burns. It is very irritating to the eyes, nose, throat, and skin. Wash affected areas with water immediately. Inhalation of large quantities can cause unconsciousness. Contact with skin causes a yellow stain that cannot be washed off.

14. Nitrogen dioxide (formed from the reaction of nitric acid and copper metal). CAUTION. The fumes are suffocating and poisonous. Evolution of this substance should occur in a fume hood. It is a severe eye, nose, throat, and skin irritant. This substance affects the lungs and respiratory system. Chest pains may occur.

15. Phosphorus pentoxide. CAUTION—POISON. This substance is a strong irritant. It is corrosive to the skin, mucous membranes, and eyes. Avoid ingestion, inhalation, and skin contact. Notify your instructor if skin contact occurs. Affected area should be irrigated immediately with copious amounts of water.

16. Arsenic pentoxide. CAUTION—POISON. Avoid ingestion, inhalation, and skin contact. Acts as a skin irritant. Wash affected areas immediately with copious amounts of water.

17. Calcium oxide. CAUTION. This chemical is a strong caustic. It may be irritating to the eyes, lungs, and respiratory system. It may also cause severe irritations to the skin and mucous membranes. Avoid ingestion, inhalation, and skin contact. Wash affected area with water immediately.

INTRODUCTION

Because it brought so much order to the study of chemistry, the development of the periodic table of the elements was one of the greatest achievements in the history of chemistry. One of the major features of the periodic table is that elements with similar chemical properties are found in the same vertical column.

Such a column of elements is called a group or family of elements.

Some Things You Will Learn by Doing This Experiment

1. You will learn how to use the *Handbook of Chemistry and Physics.*
2. You will learn what types of information the periodic table of the elements contains.
3. You will learn that elements within a group have similar chemical properties.

DISCUSSION

One of the advantages of the periodic table is that it enables chemists to predict the properties of yet-undiscovered elements. This was exactly what happened with gallium. In 1869, when Mendeleev published his periodic table, the element gallium was unknown. However, Mendeleev predicted its properties, based on what he knew to be true about the elements that appeared directly above and below gallium in the table. Mendeleev predicted the melting point, boiling point, and atomic mass of gallium, which he called eka-aluminum. Then, in 1875, the French chemist Lecoq de Boisbaudran discovered gallium while he was analyzing zinc ore. Gallium had properties identical to those predicted by Mendeleev six years earlier!

In this experiment, you'll have a chance to see the properties of elements in various groups of the periodic table. You will be able to discover for yourself whether elements within a group do indeed have similar properties. We shall also observe the compounds of elements derived from the same group, for example, SrO, MgO, CaO. In the questions at the end of the experiment, we shall look at some of the other features of the periodic table.

PART 1: THE GROUP IV-A ELEMENTS

IV-A

C
Si
Ge
Sn
Pb

You are now going to observe the elements in Group IV-A. The object of this part of the experiment is to see whether you can discover similarities and differences among elements in a group on the basis of their physical properties.

PROCEDURE

1. Label five test tubes and obtain samples of the Group IV-A elements (carbon, silicon, germanium, tin, and lead).
2. Describe the physical state, color, and consistency of each element. Record your data on the report page.
3. Determine whether any of these elements are attracted to a magnet.
4. Describe any visible similarities among these elements; record them on the report page.
5. Describe any visible differences among these elements; record them on the report page.
6. Look up the melting point, boiling point, and density of each of the elements in the *Handbook of Chemistry and Physics;* record these data on the report page.
7. Record the atomic weight of each element on the report page.

PART 2: SOME GROUP II-A METAL OXIDES

II-A

Be
Mg
Ca
Sr
Ba
Ra

You are now going to examine the compounds strontium oxide (SrO), magnesium oxide (MgO), and calcium oxide (CaO). All three of these oxides are derived from the Group II-A metals. After you have observed the three compounds and recorded any similarities or differences among them, you will determine some of the chemical properties of these three compounds.

PROCEDURE

1. Obtain a small sample of each compound (SrO, MgO, and CaO) (about the size of a pea) and place each sample in a properly labeled test tube.

2. Observe the three compounds. Describe the physical state, color, and consistency of each compound. Record these data on the report page.

3. Using the *Handbook of Chemistry and Physics,* look up the melting point and boiling point of each compound. Record these data.

4. Add about 1 mL of water to each test tube.

5. Test each solution with neutral litmus paper. Describe the results. Are the results similar for all three compounds? Record your data.

. . . .

Note: Neutral litmus paper turns blue in basic solutions and red in acidic solutions.

. . . .

6. The reactions that took place were the following:

$$SrO + H_2O = Sr(OH)_2$$

$$CaO + H_2O = Ca(OH)_2$$

$$MgO + H_2O = Mg(OH)_2$$

What are the names of these three products? Why did the litmus paper turn blue?

PART 3: SOME GROUP V-A NONMETAL OXIDES

. . . .

Note: Your instructor may want to perform this part of the experiment as a demonstration.

. . . .

V-A

N
P
As
Sb
Bi

You are now going to observe the properties of some Group V-A oxides. The compounds you will observe are nitrogen dioxide (NO_2), phosphorus pentoxide (P_2O_5), and arsenic pentoxide (As_2O_5).

. . . .

Caution: Be extremely careful with these compounds! If any of them comes into contact with your skin, wash immediately with soap and water. (If soap is not readily available, at least flood your skin with plenty of water.) In addition, approach nitrogen dioxide with *great* care. Nitrogen dioxide is a gas that you must prepare according to the instructions given below. This gas is *very* toxic and should be prepared under a hood. In fact, the entire procedure for Part 3 should be performed under a hood.

. . . .

PROCEDURE

1. Obtain two clean, dry test tubes. Label one test tube arsenic pentoxide (As_2O_5) and the other phosphorus pentoxide (P_2O_5).

2. Take both test tubes to your instructor, who will either dispense the materials for you or have you disperse the materials yourself. Take a very small sample (about the size of a pea) of each material.

3. Observe the physical state, color, and consistency of each compound. Record your observations.

4. Using the *Handbook of Chemistry and Physics,* look up the melting point and boiling point of each compound and record these data. Also record the melting point and boiling point of nitrogen dioxide gas.

5. Add about 2 mL of water to each test tube containing the compounds under investigation.

6. Test the resulting solutions with neutral litmus paper. Describe the results and record them.

7. Obtain a clean, dry test tube and fill it about one-fourth full with $6M$ nitric acid (dilute nitric acid).

. . . .

Note: Be careful with the nitric acid; even dilute solutions can burn the skin.

. . . .

8. Take your test tube to the fume hood. Also obtain a piece of copper and take it to the hood. Drop the copper into the test tube containing the nitric acid. The following reaction will occur:

$$4HNO_3 + Cu \rightarrow Cu(NO_3)_2 + 2H_2O + 2NO_2$$

Brown-
yellow gas

If a reaction doesn't start after two minutes, warm the test tube slightly.

9. Obtain a piece of moist neutral litmus paper. Hold it over the test tube (use a crucible tongs to hold the litmus paper). Describe the result. Compare this result with the results of your litmus tests on the other two compounds.

10. The reactions that took place were the following:

$$2NO_2 + H_2O = HNO_3 + HNO_2$$

$$P_2O_5 + 3H_2O = 2H_3PO_4$$

$$As_2O_5 + 3H_2O = 2H_3AsO_4$$

Name the products in each of these reactions. Why did the litmus paper turn red? Write your answers on the report page.

PART 4: SOME GROUP I-A SALTS

I-A

Li
Na
K
Rb
Cs
Fr

You are now going to observe the properties of some Group I-A salts. After you have observed each of the elements, you will describe the results of the tests that you will perform on them.

PROCEDURE

1. Obtain a small sample of each compound (LiCl, NaCl, and KCl) (about the size of a pea) and place each in a properly labeled test tube.
2. Describe the physical state, color, and consistency of each compound; record your observations.
3. Look up the melting point, boiling point, density, and formula weight of each compound in the *Handbook of Chemistry and Physics* and record these data.
4. Add about 1 mL of water to each test tube.
5. Test each solution with neutral litmus paper. Describe the results. Are the results similar for all three compounds?

    When these salts dissolve in water they ionize according to the following equations:

$$LiCl \xrightarrow{H_2O} Li^{1+} + Cl^{1-}$$

$$NaCl \xrightarrow{H_2O} Na^{1+} + Cl^{1-}$$

$$KCl \xrightarrow{H_2O} K^{1+} + Cl^{1-}$$

QUESTIONS

1. Why do elements within a group have similar chemical properties?
2. Look up the atomic weights of Si and Sn. To obtain the average, add the two weights and divide by two. Now look up the atomic weight of the element Ge, which falls between Si and Sn in the table of elements. What have you discovered? Does this lend supporting evidence to the periodicity of elements? Try this test with the elements Li and K; Ne and Kr. (Note that this doesn't work for all the elements.)
3. Using your data, describe some periodic trends that you can see among the elements.
4. Describe some additional periodic trends (see your textbook).

Report on Experiment 8

Name _____

Section _____ Date _____

Instructor _____

PART 1: THE GROUP IV-A ELEMENTS

Element	Physical State	Color	Consistency	Magnetic Properties	Melting Point	Boiling Point	Atomic Weight	Density
Carbon								
Silicon								
Germanium								
Tin								
Lead								

List the similarities among the elements.

PART 2: SOME GROUP II-A METAL OXIDES

Compound	Physical State	Color	Consistency	Melting Point	Boiling Point	Reaction to neutral litmus (after addition of water to each compound)
MgO						
CaO						
SrO						

Describe what happens to neutral litmus paper when it is dipped in each solution.

Name the three compounds formed.

Why did the litmus paper turn blue?

Report on Experiment 8

Name _____

Section _____ Date _____

Instructor _____

PART 3: SOME GROUP V-A NONMETAL OXIDES

Compound	Physical State	Color	Consistency	Melting Point	Boiling Point	Reaction to neutral litmus (after addition of water to each compound)
As_2O_5						
P_2O_5						
NO_2						

Describe what happens to neutral litmus paper when it is dipped in each solution.

Name the compounds formed.

Why did the litmus paper turn red?

PART 4: SOME GROUP I-A SALTS

Element	Physical State	Color	Consistency	Melting Point	Boiling Point	Density	Formula Weight	Reaction to neutral litmus (after addition of water to each compound)
LiCl								
NaCl								
KCl								

Describe what happens when water is added to each element.

Write the ionization reactions:

Report on Experiment 8

Name _____

Section _____ Date _____

Instructor _____

RESPONSES TO QUESTIONS

1.

2.

3.

4.

5.

Experiment 9
The Law of Definite Composition

Time: About 2 hours

Materials: Test tubes (clean and dry), potassium chlorate ($KClO_3$)

INTRODUCTION

You are now going to examine one of the most important laws in all of chemical science: the Law of Definite Composition.

This law states that the ratio by weight of elements in a compound is the same, regardless of how the compound is prepared.

This statement was proposed in 1799 by Proust, who observed that elements always combine with each other in a definite ratio by weight for a particular compound. About 50 years later another chemist, Jean Servais Stas, confirmed Proust's observations. Stas, a Belgian chemist, performed a series of quantitative experiments and found that Proust's statement held true. The Law of Definite Composition thus has stood the test of time.

Some Things You Will Learn by Doing This Experiment

1. You will be able to explain the Law of Definite Composition.

2. You will learn how this experiment demonstrates the Law of Definite Composition.

©1988 Houghton Mifflin Company

DISCUSSION

What Proust, Stas, and many other chemists have observed can be demonstrated by the following example.

When a chemist performs an experiment in which 1 g of hydrogen reacts with oxygen, the result is that exactly 8 g of oxygen are consumed. When 2 g of hydrogen are reacted with oxygen, the chemist finds that 16 g of oxygen are consumed. The chemist repeats the experiment with 3 g, 4 g, 5 g, and 6 g of hydrogen. The results are expressed in the following table.

Hydrogen Used (grams)	Oxygen Consumed (grams)	Water Produced (grams)	Ratio of H:O	Percent H in Water	Percent O in Water
1.00	8.00	9.00	1:8	11.1	88.9
2.00	16.00	18.00	1:8	11.1	88.9
3.00	24.00	27.00	1:8	11.1	88.9
4.00	32.00	36.00	1:8	11.1	88.9
5.00	40.00	45.00	1:8	11.1	88.9
6.00	48.00	54.00	1:8	11.1	88.9

The chemist uses the following procedure to calculate the percent of hydrogen and the percent of oxygen in the water.

$$\text{Percent H} = \frac{\text{wt. of hydrogen}}{\text{wt. of water}} \times 100 = \frac{1.00 \text{ g}}{9.00 \text{ g}} \times 100 = 11.1$$

$$\text{Percent O} = \frac{\text{wt. of oxygen}}{\text{wt. of water}} \times 100 = \frac{8.00 \text{ g}}{9.00 \text{ g}} \times 100 = 88.9$$

What the experiment shows is that water is *always* formed from hydrogen and oxygen when they react in the *weight ratio* of 1:8. For additional proof of the Law of Definite Composition, one can also perform the reverse experiment. One can decompose various amounts of water into hydrogen and oxygen. Regardless of the amount of water decomposed, the hydrogen-to-oxygen ratio will be 1:8.

Initial Amount of Water (grams)	Hydrogen Produced (grams)	Oxygen Produced (grams)	Mass Ratio of H:O
100.0	11.1	88.9	1:8
50.0	5.50	44.5	1:8
27.0	3.00	24.0	1:8
9.00	1.00	8.00	1:8

In this experiment you will have a chance to add your own additional proof of the Law of Definite Composition. You will decompose the compound potassium chlorate ($KClO_3$) into potassium chloride (KCl) and oxygen gas (O_2) by heating. At the beginning of the experiment, you and your partner will have different starting amounts of potassium chlorate ($KClO_3$). Each of you will obtain different *weights* of products, but the *ratio* of potassium chloride to oxygen will be the same.

PROCEDURE

1. Obtain a clean, dry test tube and record its weight to the nearest 0.01 g. Be sure your test tube is very clean.
2. Add a small amount of potassium chlorate ($KClO_3$) to the test tube; reweigh and record the total weight.
3. To find out the amount of potassium chlorate ($KClO_3$) in the test tube, subtract the weight of the test tube from the total weight.

. . . .

Note: For convenience, the sample should weigh between 1 g and 3 g. This sample size is chosen to keep within the limitations of the equipment available.

. . . .

4. If the sample is less than 1 g or more than 3 g, add or remove some potassium chlorate and reweigh the test tube.

. . . .

Note: Try to obtain samples such that you and your partner have different weights of potassium chlorate.

. . . .

5. Arrange the apparatus as shown in Figure 9-1. Make sure that you clamp the test tube near the top.
6. Heat the test tube gently. As oxygen is given off, the potassium chlorate will appear to melt.
7. After five minutes of gentle heating, heat with a strong flame.

. . . .

Note: Make *sure* that you heat all sides of the test tube equally. Don't localize the heat in one place. If you do, the test tube could melt or crack. Also, you may not succeed in decomposing all the potassium chlorate.

. . . .

Potassium chlorate

Figure 9-1
Experimental setup for heating potassium chlorate

8. Continue to heat the test tube for an additional ten minutes, to be sure that all the potassium chlorate has been decomposed.

9. Allow the test tube to cool before weighing it. The test tube now contains potassium chloride (KCl).

10. Find the weight of the potassium chloride by subtraction as shown.

>   Weight of test tube + potassium chloride
> − Weight of test tube
> ――――――――――――――――――――――
>   Weight of potassium chloride

Record this weight on the report page.

11. Obtain the weight of the oxygen released as shown.

>   Weight of test tube + potassium chlorate
> − Weight of test tube + potassium chloride
> ――――――――――――――――――――――
>   Weight of oxygen released.

Record this weight.

12. From your data, calculate the ratio of potassium chloride to oxygen. Do the same using your partner's data. How do your results compare with your partner's?

QUESTIONS

1. You can calculate the theoretical ratio of potassium chloride to oxygen from the formula for potassium chlorate ($KClO_3$). In order to do this, you must obtain the atomic weights of the elements from the periodic table. You then determine what *percentage* of the formula weight the potassium chloride contributes to the potassium chlorate and then determine what *percentage* the oxygen contributes to the potassium chlorate. Using these values, you then determine the theoretical ratio of potassium chloride to oxygen.

Theoretical ratio of potassium chloride to oxygen

$$= \frac{\text{percentage potassium chloride}}{\text{percentage oxygen}}$$

(You obtain the percentage of potassium chloride and the percentage of oxygen from the formula for $KClO_3$.) Determine this theoretical ratio, and then determine the percentage error in your results as compared to the theoretical values. The formula for percentage error is

$$\text{Percentage error} = \frac{|\text{theoretical value} - \text{experimental value}|}{\text{theoretical value}} \times 100$$

The "bars" in the numerator of the formula mean "absolute value." This means that we're interested in the difference between the theoretical value and the experimental value, not in whether the difference is positive or negative.

2. How could you determine experimentally whether all the potassium chlorate ($KClO_3$) had been decomposed?

3. Why must the test tube be cool before you weigh it?

4. What reasons can you offer for any differences between your results and your partner's results?

Report on Experiment 9

Name _____

Section _____ Date _____

Instructor _____

	Quantity	How to Obtain	Your Values	Partner's Values
1.	Wt. of test tube	Weigh		
2.	Wt. of test tube plus potassium chlorate	Weigh		
3.	Wt. of potassium chlorate	(2) − (1)		
4.	Wt. of test tube plus potassium chloride	Weigh		
5.	Wt. of potassium chloride	(4) − (1)		
6.	Wt. of oxygen	(2) − (4)		
7.	Ratio of potassium chloride to oxygen	(5) ÷ (6)		

CALCULATIONS

RESPONSES TO QUESTIONS

1.

2.

3.

4.

Experiment 10
Types of Chemical Reactions

Time: About 2 hours

Materials: Iron dust, sulfur powder, mercury(II) oxide (HgO), mossy zinc, $6M$ HCl, $0.5M$ $Pb(NO_3)_2$ solution, $1M$ KI solution

Safety and Chemical Hazard Guide

Please review this before performing the experiment.

1. Glasses must be worn at *all* times when in the laboratory!

2. Iron dust. No special hazards. However, avoid breathing dust.

3. Sulfur powder. Causes irritation of the skin and mucous membranes. Avoid ingestion and inhalation.

4. Mercury(II) oxide. CAUTION. Avoid inhalation, ingestion, and skin contact or mercury poisoning may result. Wash skin immediately if contact occurs. The target organs of this substance are the skin, respiratory system, central nervous system, and kidneys.

5. Zinc metal. Inhalation of fumes may result in sweet taste, throat dryness, cough, weakness, generalized aching, chills, fever, nausea, and vomiting. Avoid inhalation of this substance.

6. Hydrochloric acid ($6M$). CAUTION. This chemical can cause severe burns on skin, as well as in eyes, nose, and throat. Coughing, choking, and skin rashes can result. Avoid inhalation, ingestion, and skin contact. Wash affected area immediately with plenty of water. Notify instructor.

7. Lead(II) nitrate ($0.5M$). Avoid ingestion, which can cause lead poisoning. Avoid contact with skin. If contact is made, wash affected area immediately with soap and water. Wash hands after handling solution. Lead solutions affect the central nervous system, gastrointestinal tract, kidneys, and blood.

8. Potassium iodide ($1M$). No special hazards. However, avoid ingestion.

INTRODUCTION

There are thousands of chemical reactions and many ways to classify them. For a course in basic chemistry, we can simplify matters by learning four types of chemical reactions.

1. Combination reactions

2. Decomposition reactions

3. Single-replacement reactions

4. Double-replacement reactions

Here you will have the chance to perform experiments involving each of these types of reactions.

Some Things You Will Lean by Doing This Experiment

1. You will learn how to carry out various chemical reactions you have read about in your textbook.

2. You will get a chance to use the various lab techniques you have been acquiring in your previous lab work.

3. You will learn the names of the four types of reactions performed in this experiment, and be able to give an example of each.

4. Given a chemical equation, you will learn how to determine whether it is one of the four types of reactions discussed in this experiment and which type it is.

DISCUSSION

PART 1: COMBINATION REACTIONS

Combination reactions are those in which two or more substances combine to form a more complex substance. The general formula for such a reaction is

$$A + B \rightarrow AB$$

Many combination reactions involve the union of two elements to form a compound.

Fe	+ S	$\xrightarrow{\text{heat}}$	FeS
Iron	Sulfur		Iron(II) sulfide

C	+ O_2	$\xrightarrow{\text{heat}}$	CO_2
Carbon	Oxygen gas		Carbon dioxide

H_2	+ Cl_2	$\rightarrow$	$2HCl$
Hydrogen gas	Chlorine gas		Hydrogen chloride

$2Mg$	+ O_2	$\xrightarrow{\text{heat}}$	$2MgO$
Magnesium	Oxygen gas		Magnesium oxide

The last reaction was performed in Experiment 5.

As an example of a combination reaction, you are going to observe the reaction of iron and sulfur.

PROCEDURE

1. Using a balance, weigh 7 g of iron powder.

. . . .

Note: Don't place the iron powder directly on the balance pan. Use a weighing cup!

. . . .

2. Next, weigh 4 g of sulfur powder. Place the sulfur powder in a cup also.

3. Obtain a mortar and pestle and pour the weighed portions of iron and sulfur into the mortar. Grind the iron and sulfur together until you have mixed them together fairly well.

4. Transfer the iron-sulfur mixture to a test tube.

5. Take the mixture to the fume hood and, using a test-tube holder to hold the test tube, heat the mixture with a gas burner. In about one minute the mixture will begin to glow. This is the signal that a reaction has taken place. At this point, *stop heating* and allow the test tube to cool.

6. The material in the test tube is no longer a mixture of iron and sulfur; it is the compound iron(II) sulfide.

7. You will now show that the compound has chemical properties different from those of the mixture. Place a small amount of sulfur (about the size of a pea) in a clean, dry test tube.

8. Place a small amount of iron dust (about the size of a pea) in another clean, dry test tube.

9. Take the test tubes from Steps 7 and 8 to the fume hood, along with the test tube containing the iron(II) sulfide.

10. Using an eye dropper, add 10 drops of 6*M* HCl (dilute HCl) to the test tube containing the iron(II) sulfide. What do you smell?

. . . .

Caution: Use the proper technique for smelling the gas being generated. It is highly poisonous! The proper technique for smelling a substance involves holding the material about a *foot away from your nose* and wafting the odor toward you with your hand. *Do not* put your nose over the test tube containing the material!

. . . .

11. Add 10 drops of 6*M* HCl to the test tube containing the iron powder. Next add 10 drops of 6*M* HCl to the test tube containing the sulfur powder. Did you get the same results in this step as you did in Step 10?

12. Try adding a few drops of HCl to a *mixture* of iron and sulfur. What are the results?

What does this experiment tell you about the properties of iron and sulfur compared with those of iron(II) sulfide?

PART 2: DECOMPOSITION REACTIONS

A decomposition reaction is the reverse of a combination reaction. It involves the breakdown of a complex substance into simpler substances. The general formula for this type of reaction is

$$AB \rightarrow A + B$$

Some examples of decomposition reactions are

$$2HgO \xrightarrow{\text{heat}} 2Hg + O_2$$

Mercury(II)   Mercury   Oxygen
oxide   metal   gas

$$2H_2O \xrightarrow{\text{electricity}} 2H_2 + O_2$$

$$2KClO_3 \xrightarrow{\text{heat}} 2KCl + 3O_2$$

Potassium   Potassium   Oxygen
chlorate   chloride   gas

The last reaction was performed in Experiment 9.

PROCEDURE

1. Obtain a wooden splint.
2. Obtain a clean, dry test tube and record its weight.
3. Add about 1 g of HgO, which is mercury(II) oxide (also called mercuric oxide), to the test tube and reweigh.
4. Determine the amount of mercury(II) oxide by subtraction. If the amount of mercury(II) oxide is much greater or much less than 1 g, remove or add material and reweigh.
5. After you have obtained about 1 g of mercury(II) oxide in your test tube, you are ready to decompose it.
6. Hold the test tube with a test-tube holder and heat the mercury(II) oxide in the hottest part of your burner flame.

. . . .

Caution: Do not let the test tube point at anyone. Also, do not inhale any of the vapor escaping from the test tube.

. . . .

7. After one minute of heating, light the wooden splint by holding it in the flame of your gas burner. After the wooden splint has burned for a few seconds, blow the flame out. The splint should continue to glow.
8. Place the glowing splint inside the test tube containing the mercury(II) oxide, which you are still heating.
9. If the compound is giving off oxygen, the glowing splint will ignite.
10. Stop heating the mercury(II) oxide and look carefully at the test tube. Do the sides of the test tube have a shiny metallic coating? What substance do you think this is?
11. Ask your instructor how you should dispose of the test tube and its contents. After all, we don't want mercury compounds being poured down the sink and ending up in a nearby river or lake.

PART 3: SINGLE-REPLACEMENT REACTIONS

A single-replacement reaction is one in which one element replaces another element from its compound. The general formula for this type of reaction is

$$A + BC \rightarrow AC + B$$

Some examples of single-replacement reactions are

Zn	+ 2HCl	→ ZnCl$_2$	+ H$_2$
Zinc metal	Hydrochloric acid	Zinc chloride	Hydrogen gas

Mg	+ Cu(NO$_3$)$_2$	→ Mg(NO$_3$)$_2$	+ Cu
Magnesium metal	Copper(II) nitrate	Magnesium nitrate	Copper metal

2Na	+ 2H$_2$O	→ 2NaOH	+ H$_2$
Sodium metal	Water	Sodium hydroxide	Hydrogen gas

As an example of a single-replacement reaction, you're going to observe the reaction of zinc metal with hydrochloric acid. You'll be able to check on the reaction by testing for the evolution of hydrogen gas.

PROCEDURE

1. Place one or two pieces of mossy zinc into a clean test tube.

. . . .

Note: Mossy zinc is an impure form of zinc that has been heated to molten form and poured drop by drop into water.

. . . .

2. Pour about 10 mL of 6*M* HCl into a graduated cylinder.

3. Pour the 6*M* HCl into the test tube containing the mossy zinc.

4. Immediately invert a second test tube over the first one as shown in Figure 10-1. If hydrogen gas is being generated, it will collect in the inverted test tube.

5. After the reaction has ceased, quickly remove the inverted test tube, holding it with its mouth down, and place the mouth of the test tube directly in the flame of your gas burner. If hydrogen gas is present, you'll hear a barking or popping sound as the hydrogen gas burns with oxygen in the air.

6. Pour the contents of the first test tube into an evaporating dish and take the evaporating dish to a fume hood. Heat the dish (in the fume hood) until all the liquid has evaporated.

7. What substance is left in the dish after the liquid has evaporated?

PART 4: DOUBLE-REPLACEMENT REACTIONS

A double-replacement reaction is one in which two compounds exchange ions with each other. Double-replacement reactions are usually carried out in a water solution. The general formula for this type of reaction is

$$AC + BD \rightarrow AD + BC$$

Hydrogen gas
← collected in
this test tube

This test tube
← contains 6M̲HCl
and mossy zinc

Figure 10-1
Collecting hydrogen gas

Some examples of double-replacement reactions are

$$2KI \ (aq) \quad + Pb(NO_3)_2 \ (aq) \rightarrow 2KNO_3 \ (aq) + PbI_2 \ (s)$$

| Potassium iodide | Lead(II) nitrate | Potassium nitrate | Lead(II) iodide |

$$AgNO_3 \ (aq) + KCl \ (aq) \qquad \rightarrow KNO_3 \ (aq) \quad + AgCl \ (s)$$

| Silver nitrate | Potassium chloride | Potassium nitrate | Silver chloride |

$$HCl \ (aq) \quad + NaOH \ (aq) \quad \rightarrow NaCl \ (aq) \quad + H_2O$$

| Hydrochloric acid | Sodium hydroxide | Sodium chloride | Water |

In these examples, the notation (*aq*) means that the substance is soluble in water (*aqua* is Latin for "water"); in other words, the substance exists as ions in solution. The notation (*s*), for "solid," means that the substance is insoluble in water; in other words, it precipitates out of solution as a solid.

As an example of a double-replacement reaction, you're going to observe the reaction of potassium iodide with lead nitrate. Since lead iodide separates from solution, you'll be able to filter it out. You may then obtain the powdered potassium nitrate by evaporating the water from the filtrate.

PROCEDURE

1. Place 10 mL of 1*M* KCl (potassium iodide) solution in a clean test tube.

2. Place 10 mL of 0.5*M* Pb(NO₃)₂ [lead(II) nitrate] solution in another test tube.

3. Both potassium iodide and lead(II) nitrate are white crystalline salts that are soluble in water. Keep this in mind as you continue the experiment.

4. Mix the contents of the two test tubes together. What happens?

5. The equation for the reaction is

   $$2KI\ (aq) + Pb(NO_3)_2\ (aq) \rightarrow 2KNO_3\ (aq) + PbI_2\ (s)$$

   Yellow

6. Filter the PbI_2 from the test tube.
7. Collect the filtrate in an evaporating dish and boil off the water. What compound is left in the dish?

QUESTIONS

1. Classify each of the following reactions as either a combination reaction, a decomposition reaction, a single-replacement reaction, or a double-replacement reaction.

   a) $Mg + 2HCl \rightarrow MgCl_2 + H_2$

   b) $2NaOH + H_2SO_4 \rightarrow Na_2SO_4 + 2H_2O$

   c) $2NaCl \rightarrow 2Na + Cl_2$

   d) $C + O_2 \rightarrow CO_2$

   e) $SO_3 + H_2O \rightarrow H_2SO_4$

2. In the procedure for Part 3, you were asked to heat the evaporating dish under the hood (Step 6 of the procedure). Why should this step be carried out under a hood?

Report on Experiment 10

Name _____

Section _____ Date _____

Instructor _____

PART 1: COMBINATION REACTIONS

1. Description of reaction between iron and sulfur:

2. Odor observed when you added HCl to

    a) iron(II) sulfide

    b) iron

    c) sulfur

    d) mixture of iron powder and sulfur powder

PART 2: DECOMPOSITION REACTIONS

1. Description of decomposition of mercury(II) oxide:

2. Description (and results) of the test for oxygen with the glowing splint:

3. The substance coating the sides of the test tube is _____ .

PART 3: SINGLE-REPLACEMENT REACTIONS

1. Description of reaction between zinc and hydrochloric acid:

2. Test for hydrogen (positive or negative):

3. Name of the substance left after liquid evaporated from evaporating dish:

PART 4: DOUBLE-REPLACEMENT REACTIONS

1. Description of the reaction between potassium iodide and lead(II) nitrate:

2. Name of the substance left in the evaporating dish after the water is evaporated:

Write balanced equations for each of the reactions you carried out in this experiment.

1.

2.

3.

4.

Report on Experiment 10

Name _____

Section _____ Date _____

Instructor _____

RESPONSES TO QUESTIONS

1. a)

    b)

    c)

    d)

    e)

2.

Experiment 11
Determination of the Amount of Phosphate in Water

Time: About 2 hours

Materials: Colorimeter, standard phosphate solutions, unknown phosphate solutions, color reagent, 100-mL beakers, 25-mL graduated cylinders, 5-mL graduated cylinders

. . . .

Safety and Chemical Hazard Guide

Please review this before performing the experiment.

1. Glasses must be worn at *all* times when in the laboratory!

2. Stock phosphate solution (potassium dihydrogen phosphate, 100 mg/L). No special hazards.

3. Color reagent (contains sulfuric acid, potassium antimonyl tartrate, ammonium molybdate, and ascorbic acid). CAUTION. This substance contains sulfuric acid, which can cause burns upon contact. Avoid ingestion, inhalation, and skin contact. Wash affected area immediately with water.

INTRODUCTION

Phosphates exist in nature in many forms. In the past few years, industry has begun to add significantly to the total. Although the controversy about the use of phosphates in detergents and the effect of phosphates on the ecology of streams, lakes, and ponds continues, it is still important to know how much phosphate a body of water contains. There is one thing that ecologists know for sure: when phosphates are present in large amounts, they do affect the ecology of a stream or lake. This is because the phosphates act as a fertilizer or food for the plant life in the stream.

> *Given this fertilizer, plants are able to multiply so rapidly that they fill in the stream or lake. This process is called eutrophication. A high percentage of phosphates in drinking water gives the water a soapy, bubbly taste.*

In this experiment you will determine the amount of phosphate in a sample of water.

Some Things You Will Learn by Doing This Experiment

1. You will learn how to operate a colorimeter.

2. You will learn the principle behind colorimetric determinations.

3. You will learn how to plot a graph of absorbance versus concentration, and you will learn how to use this graph to find the concentration of an unknown sample when you know its absorbance.

DISCUSSION

To determine the amount of phosphate in a sample of water, you'll use the colorimetric method. The principle behind this method is to find a chemical agent (or agents) that reacts specifically with phosphates, so that a colored solution results. The intensity of the color depends on the amount of phosphate in the sample. Such chemical agents do exist for phosphates. Your instructor may wish to discuss the reagents involved in this method of determining phosphate concentrations in water.

The colorimetric procedure involves making up a series of known phosphate solutions and treating them with a chemical agent (called the color reagent). At the same time, one also treats the water sample (whose phosphate concentration is to be determined) with the same chemical agent. After the color develops in the solutions, one compares the color of the water sample with the color of the known phosphate solutions.

A person can, of course, compare colors by visual observation, but this method is not always reliable, especially if the intensity of the colors is very similar. A much better method is to use a device called a *colorimeter,* an instrument that measures the intensity of the color electronically. The colorimeter does this by measuring the absorbance of a specific wavelength of light as it passes through the colored sample.

Each of the known phosphate solutions absorbs a different amount of light. The more phosphate in the solution, the more light it absorbs. The observer reads the absorbance of each sample on the colorimeter meter and then draws a graph, plotting the concentration of each *known* phosphate solution against its absorbance. The procedure should give a straight-line graph. This graph enables one to determine the phosphate content of the water sample, by simply finding the absorbance of the sample on the graph and seeing what concentration of phosphate this absorbance corresponds to.

In our description of this procedure, we are assuming that solutions of all of the known phosphates have been prepared, and that the color reagent that forms complexes with phosphates has been prepared also.

PART 1: CALIBRATION OF COLORIMETER

PROCEDURE

(The procedure here was written for use with the Bausch and Lomb Spectronic® 20 spectrophotometer shown in Figure 11-1. If you are using another type of colorimeter, some minor changes in the procedure might be needed.)

1. Obtain your colorimeter and colorimeter test tubes.

2. Turn on the instrument and adjust the wavelength-control knob to 660 nanometers. Allow the instrument to warm up for 10 minutes.

1—Pilot light   5—Percent-transmission knob
2—Meter   6—Zero-control knob
3—Wavelength control   7—Sample holder
4—Wavelength scale

Figure 11-1
The Spectronic® 20 spectrophotometer

3. With the sample compartment empty, adjust the zero-control knob (the left-hand knob) so that the meter pointer reads infinite absorbance. (This is the same as zero-percent transmittance.) Close the cover of the sample holder when you are adjusting the meter.

4. Fill one colorimeter test tube with distilled water and place it in the sample compartment of the instrument. Close the cover. The meter pointer should move to the right.

5. Using the percent-transmittance knob (the right-hand knob), adjust the meter pointer to read zero absorbance. (This is the same as 100-percent transmittance.)

6. The instrument is now calibrated for use. You may wish to repeat the above procedure to be sure that the instrument is stable.

7. If you wish to learn more about the operation of the colorimeter, ask your instructor for a copy of the instrument manual.

PART 2: DETERMINATION OF PHOSPHATE

PROCEDURE

1. Obtain from your instructor the standard phosphate solutions of known concentration. Also obtain your unknown phosphate solution.

2. Prepare these solutions in the following manner.

   a) Using your 25-mL graduated cylinder, carefully measure 25 mL of each known solution. Do the same for the unknown solution. Pour each solution into its own 100-mL beaker.

   b) Using your small graduated cylinder, measure 5 mL of color reagent for each solution. Add this color reagent to each solution in its beaker.

   c) Stir each solution and allow the color to develop for 15 minutes.

3. Place each solution in its own colorimeter test tube and measure its absorbance in the colorimeter. Close the cover of the sample holder when you are taking a reading.

4. Construct a graph (use page 100 of this manual). Plot absorbance versus concentration for each of the known phosphate samples. From this graph and the absorbance of your unknown sample, determine the concentration of phosphate in your unknown sample.

QUESTIONS

1. The type of phosphates dealt with in this experiment are called *orthophosphates*. These are the type that have the $(PO_4)^{3-}$ ion. What is the name of the type of phosphate found in detergents? What is the general formula for this type of phosphate? (*Hint:* To answer this question, you might look on a box of phosphate-containing detergent or use a reference source in your library.)

2. What are the chief sources of phosphates in the nation's waterways, lakes, and water supplies?

3. What could you do to obtain the concentration of phosphate in your unknown sample if the following situation occurred?

   You measure out 25 mL of your unknown solution. You add to your sample 5 mL of color reagent. The color develops. After 15 minutes you try to read its absorbance on the colorimeter. However, you find that the color of the solution is so intense that it registers off the meter at infinite absorbance.

Report on Experiment 11

Name _____

Section _____ Date _____

Instructor _____

Concentration of Known Phosphate Solution, in parts per million (ppm)	Absorbance of Known Phosphate Solution

Unknown number _____

Absorbance of unknown _____

Concentration of unknown obtained from graph _____ ppm.

RESPONSES TO QUESTIONS

1.

2.

3.

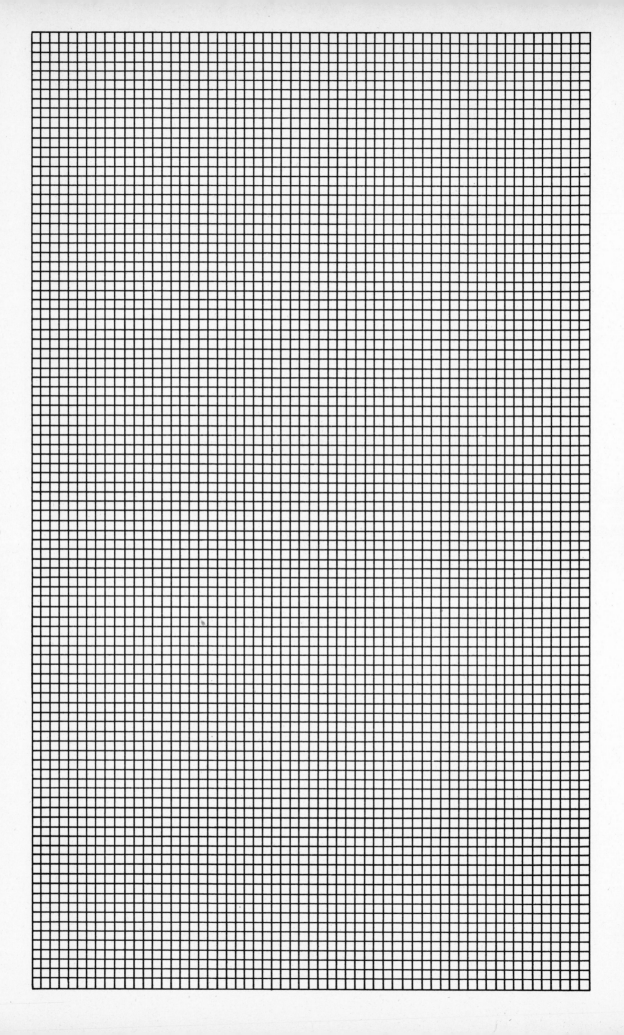

Experiment 12
Measuring the pH of Some Acids, Bases, and Salts

Time: About 2 hours

Materials: pH meter, pH paper (or universal indicator), 1M ammonium chloride, 1M potassium chloride, 1M sodium carbonate, 0.1M sodium hydroxide, 0.1M hydrochloric acid, 0.1M ammonium hydroxide, vinegar (or 4-percent-by-weight acetic acid solution), orange juice, milk, a soft drink, buffer solution (for standardization of pH meter)

Safety and Chemical Hazard Guide

Please review this before performing the experiment.

1. Glasses must be worn at *all* times when in the laboratory!
2. Ammonium chloride (1M). No special hazards. However, avoid ingestion and skin contact.
3. Potassium chloride (1M). No special hazards. However, large doses by mouth can cause gastrointestinal irritation, vomiting, weakness, and circulatory disturbances.
4. Sodium carbonate (1M). CAUTION. This solution is alkaline and may cause burns on contact. Sensitivity reactions may occur from repeated skin contact. Ingestion of large quantities may produce corrosion of gastrointestinal tract, vomiting, diarrhea, and circulatory collapse. Avoid ingestion, inhalation, and skin contact. Wash affected area immediately with water.
5. Sodium hydroxide (0.1M). CAUTION. This chemical can cause skin burns on contact. Wash affected area immediately with plenty of water and notify your instructor. This chemical can also cause eye, nose, and throat burns and temporary loss of hair on skin upon contact.
6. Hydrochloric acid (0.1M). CAUTION. This chemical can cause severe burns on the skin and in the eyes, nose, and throat. Coughing, choking, and skin rashes can result. Avoid inhalation, ingestion, and skin contact. Wash affected area immediately with plenty of water and notify your instructor.

7. Ammonium hydroxide (0.1M). CAUTION. This chemical can cause skin burns on contact. Wash affected area immediately with plenty of water and notify your instructor. This substance is also an eye, nose, and throat irritant. Avoid inhalation and ingestion. Inhalation of this substance can cause bronchial spasms (coughing), chest pain, and pulmonary edema. Target organs are the eyes, lungs, and respiratory system.

8. Vinegar (5% acetic acid solution). Avoid splashing into eyes or on open cuts. Although vinegar presents no special hazards, more concentrated solutions of acetic acid should be handled with extreme caution. Concentrated solutions of acetic acid can cause asphyxiation, nausea, vomiting, chest pains, weakness, stupor, convulsions, and eye irritations. Avoid inhalation, ingestion, and skin contact.

INTRODUCTION

The pH of a substance reflects the degree to which it is acidic or basic.

In this experiment you are going to measure the pH of various substances. The pH scale is numbered from 0 to 14. Figure 12-1 shows the range of the pH scale.

A solution with a pH between 0 and 2 is considered very acidic. A solution with a pH between 2 and 5 is moderately acidic, and one with a pH between 5 and 7 is slightly acidic. A solution whose pH is equal to 7 is neutral. If a solution has a pH above 7, it is basic. A solution with a pH between 7 and 9 is slightly basic, one with a pH between 9 and 11 is moderately basic, and one with a pH between 11 and 14 is very basic. Table 12-1 lists the pH of some common substances.

Some Things You Will Learn by Doing This Experiment

1. You will learn that the pH scale is a means of measuring the acidity or basicity of a substance.
2. You will learn that there are different ways to measure pH.
3. You will learn that if the pH of a substance is *below* 7, it is acidic; if the pH is *above* 7, the substance is basic; and if the pH of a substance is 7, it is neutral.

DISCUSSION

There are many ways to determine the pH of a solution. A simple method involves placing a few drops of a chemical dye (also called a chemical indicator) into the

0 1 2 3 4 5 6 7 8 9 10 11 12 13 14

Very acidic   Moderately acidic   Slightly acidic   Neutral   Slightly basic   Moderately basic   Very basic

Figure 12-1
The pH scale

Table 12-1
The pH of Some Common Substances

Substance	pH
1M HCl	0
Lemons	2.3
Vinegar	2.8
Soft drinks	3.0
Oranges	3.5
Tomatoes	4.2
Rainwater	6.2
Milk	6.5
Pure water	7.0
Sea water	8.5
$Mg(OH)_2$ (saturated solution)	10.5
0.1M NH_4OH	11.1
1M NaOH	14.0

solution to be tested. The dye turns a specific color, depending on the pH of the solution.

Another method of determining pH involves using an instrument called a pH meter, which measures the pH of a solution electronically. The meter is standardized, before each use, with solutions of known pH. Once the pH meter is standardized, one can use it to obtain the pH of unknown solutions.

In this experiment we shall use both the indicator and pH meter methods to determine the pH of various substances.

PART 1: DETERMINATION OF pH USING UNIVERSAL INDICATOR (OR pH PAPER)

PROCEDURE

1. Obtain a test tube rack and 9 clean, dry test tubes.
2. Label each of the test tubes as follows:
    a) 1M ammonium chloride, NH_4Cl
    b) 1M potassium chloride, KCl
    c) 1M sodium carbonate, Na_2CO_3
    d) 0.1M ammonium hydroxide, NH_4OH
    e) Vinegar
    f) Orange juice
    g) Milk
    h) Soft drink
    i) Tap water
3. Obtain about 2 mL of each substance and pour it into the properly labeled test tube.
4. If you are using *universal indicator,* use the following procedure:
    a) Add a few drops of the indicator to each test tube. The instructions on the indicator bottle (or your instructor) will tell you how much indicator to add for a 2-mL sample.
    b) Shake each test tube to mix the indicator with the solution.
    c) To obtain the pH of the solution, compare the color of each solution with the color chart supplied with the indicator.

©1988 Houghton Mifflin Company

5. If you are using pH paper, use the following procedure:
   a) Obtain a small piece of pH paper from its dispenser.
   b) Dip the stirring rod into a test tube containing one of the solutions you want to test and transfer a drop of the solution to the pH paper. Be sure you *just moisten* the paper with the solution. If you wet the paper too much, you'll simply pull all the dye out of the paper.
   c) To determine the pH of the solution, match the color of the moistened paper with the color chart on the pH dispenser.
6. Using either the method described in Step 4 or Step 5, complete the following steps:
   a) Into a clean, dry test tube measure 10 mL of 0.1M hydrochloric acid and obtain its pH.
   b) Into another clean, dry test tube measure 10 mL of 0.1M sodium hydroxide and obtain its pH.
   c) Mix the contents of the two test tubes and obtain the pH of the resultant solution.
   d) Write the chemical reaction for what occurred when the two solutions were mixed.

PART 2: DETERMINATION OF pH USING A pH METER

(This part of the experiment may be done as a demonstration by your instructor.)

PROCEDURE

1. Obtain a pH meter, plug it into the wall socket, and allow it to warm up for about 5 minutes.
2. Obtain 5 clean, dry 150-mL beakers and label them as follows:
   a) Buffer solution (pH = 7 recommended)
   b) Tap water
   c) Vinegar
   d) Milk
   e) Orange juice
3. Fill each beaker half full with the material listed on the label.
4. Because there are many types of pH meters, your instructor will give you detailed directions on how to use your pH meter. However, there are some general rules to follow with every pH meter.
   a) Handle the electrodes carefully when you place them in the beaker.
   b) Make sure that the tips of the electrodes are submerged in the solution whose pH you are measuring.
   c) Be sure that the electrodes are not touching each other or the sides of the beaker.
   d) Rinse the electrodes with distilled water between pH measurements.
5. The pH meter must be standardized before you use it.
   a) Place the clean electrodes in the buffer solution (whose pH is known).
   b) Press the "pH read" switch. This causes the pointer on the pH meter to move.

    c) Use the "calibrate" switch to adjust the pointer to the pH of the buffer solution.

6. Clean the electrodes and place them in the sample of tap water. Press the "pH read" switch; when the pointer comes to rest, read the pH of the tap water from the meter. Turn the pH switch to "standby" and remove the water sample.

7. Repeat Step 6 for each of the other solutions.

QUESTIONS

1. What are some of the advantages of using a pH meter compared with using pH paper (or universal indicator) to determine the pH of a substance?

2. What are some of the advantages of using pH paper (or universal indicator) compared with using a pH meter?

3. The pH scale is a logarithmic scale. This means that the difference between each pH unit is really a factor of ten (in terms of actual magnitude of the acidity or basicity of the substance). If the pH of grapefruit juice is 3 (pH = 3), and the pH of beer is 5 (pH = 5), how many times more acidic is the grapefruit juice than the beer?

Report on Experiment 12

Name _____

Section _____ Date _____

Instructor _____

PART 1: DETERMINATION OF pH USING UNIVERSAL INDICATOR (OR pH PAPER)

Substance	pH
1M NH$_4$Cl	
1M KCl	
1M Na$_2$CO$_3$	
0.1M NH$_4$OH	
Vinegar	
Orange juice	
Milk	
Soft drink	
Tap water	
0.1M HCl	
0.1M NaOH	
Solution of HCl and NaOH	

Write the chemical equation for the reaction between HCl and NaOH.

PART 2: DETERMINATION OF pH USING A pH METER

Substance	pH
Tap Water	
Vinegar	
Milk	
Orange juice	

RESPONSES TO QUESTIONS

1.

2.

3.

Experiment 13
Water of Hydration: The Formula for a Hydrate

Time: About 2 hours

Materials: Various hydrates for unknown samples

. . . .

Safety and Chemical Hazard Guide

Please review this before performing the experiment.

1. Glasses must be worn at *all* times when in the laboratory!
2. Various hydrated salts. No special hazards. However, since these salts are being given to you as unknowns, avoid ingestion, inhalation, and skin contact. If skin contact occurs, wash the affected area immediately with water. Notify your instructor.

INTRODUCTION

In this experiment you are going to examine a class of salts known as *hydrates*. Before we define hydrate, let's review our definition of a salt. Salts are compounds composed of a metal ion plus a nonmetal (or polyatomic) ion. Some examples are $NaCl$, $AgBr$, K_2SO_4, and $FePO_4$. Salts may also be composed of a polyatomic ion that has a positive oxidation number plus a nonmetal (or polyatomic) ion that has a negative oxidation number. Examples of these salts are NH_4Cl and NH_4NO_3.

Now let us define hydrated salts.

They are salts that combine with water to form crystalline compounds called hydrates.

In other words, water becomes part of the salt crystal. Some common hydrates are

$Na_2B_4O_7 \cdot 10H_2O$	Borax (or sodium tetraborate decahydrate)
$CaSO_4 \cdot 2H_2O$	Gypsum (or calcium sulfate dihydrate)
$KAl(SO_4)_2 \cdot 12H_2O$	Alum (or potassium aluminum sulfate dodecahydrate)
$MgSO_4 \cdot 7H_2O$	Epsom salts (or magnesium sulfate heptahydrate)

$$Na_2CO_3 \cdot 10H_2O \qquad \text{Washing soda (or sodium carbonate decahydrate)}$$

$$CuSO_4 \cdot 5H_2O \qquad \text{Blue vitriol (or copper sulfate pentahydrate)}$$

The centered dot between the formula for the salt and the formula for the water is not a decimal point, nor does it mean "multiply by." It is a way of showing that there is a *bond* between the salt and the water. When such salts are heated, this bond is *usually* broken and the water is driven off. For example:

$$MgSO_4 \cdot 7H_2O \xrightarrow{\text{heat}} MgSO_4 + 7H_2O$$

$$CuSO_4 \cdot 5H_2O \xrightarrow{\text{heat}} CuSO_4 + 5H_2O$$

A blue   A white
compound   compound

In each of these compounds, there is a definite amount of water attached to the salt crystal. To simplify matters, we may say that in $MgSO_4 \cdot 7H_2O$, for example, there are 7 water molecules for every 1 molecule of hydrated salt.

Some Things You Will Learn by Doing This Experiment

1. You will learn that some salts contain water of hydration in their crystal structures. These substances are called *hydrated salts*.
2. You will learn the procedure for determining this water of hydration.
3. You will learn how to calculate the water of hydration, using data similar to that given in this experiment.
4. You will learn that an *anhydrous salt* contains *no* water of hydration.

DISCUSSION

In this experiment you will be given a hydrated salt, and you will be told the molecular weight of the salt in its *anhydrous* form (*anhydrous* means "without water"). Using only this information, you will determine the water of hydration of your salt. Let's take some real laboratory data and go through the procedure.

An instructor gives the students a compound and says, "Determine its water of hydration. The molecular weight of this salt in its anhydrous form is 120.3. Use between 2 g and 3 g of compound for your determination, and record all weights to the nearest 0.01 g." The students weigh exactly 2.25 g of the compound into a test tube and heat it with a gas burner. After sufficient heating, the students find that the material in the test tube weighs 1.10 g. In other words, heating drove 1.15 g of water from the compound. We can represent these data as follows:

$$MX \cdot nH_2O \xrightarrow{\text{heat}} MX \qquad + nH_2O$$

2.25 g   1.10 g of   1.15 g
of hydrate   anhydrous   of water
               salt

where $MX \cdot nH_2O$ is the hydrated salt, MX is the anhydrous salt, and n represents the number of molecules of water per molecule of hydrated salt.

Although the students don't know the formula for the anhydrous salt, they can use the salt's molecular weight to determine the number of moles of anhydrous salt in 1.10 g. At the same time, they can determine the number of moles of water driven from the hydrated salt.

$$\text{Moles of MX} = (1.10 \text{ g}) \left(\frac{1 \text{ mole}}{120.3 \text{ g}} \right) = 0.00914$$

$$\text{Moles of H}_2\text{O} = (1.15 \text{ g}) \left(\frac{1 \text{ mole}}{18.0 \text{ g}} \right) = 0.0639$$

This tells the students that the ratio of MX to H_2O is 0.00914 mole to 0.0639 mole. However, we would like to report this ratio in terms of whole numbers. There are various mathematical ways of doing this. We can start out by saying:

$$\frac{\text{moles of MX}}{\text{moles of H}_2\text{O}} = \frac{0.00914}{0.0639}$$

We can simplify this ratio by dividing each number by the smaller of the two numbers, which in this instance is 0.00914. If we do this, we obtain the following:

$$\frac{0.00914}{0.00914} = 1 \quad \text{and} \quad \frac{0.0639}{0.00914} = 7$$

In other words, we can restate the previous ratio as follows:

$$\frac{\text{moles of MX}}{\text{moles of H}_2\text{O}} = \frac{1}{7}$$

Therefore the formula for the hydrate is $MX \cdot 7H_2O$. In performing your analysis, you will follow a procedure similar to that of our hypothetical students.

PROCEDURE FOR DETERMINING THE WATER OF HYDRATION

1. Weigh a clean, dry test tube. (Record all weights to the nearest 0.01 g).
2. Obtain your unknown hydrate from your instructor and record its number and molecular weight on the report page.
3. Weigh between 2 g and 3 g of hydrate directly into the test tube. Record all data on the report page.
4. Clamp your test tube to a ring stand and heat it with your gas burner for about 15 minutes. Heat the test tube uniformly. Do not concentrate all the heat at the bottom of the test tube. Heat gently at first and then more strongly. (Clamp the test tube near its mouth, as shown in Figure 9-1.)
5. Allow the test tube to cool for 10 minutes and weigh it. Record this weight on the report page.
6. Clamp the test tube back into position and heat for an additional 15 minutes.
7. Allow the test tube to cool again and reweigh it. If it has lost more than 0.05 g since the previous weighing, repeat the heating process. You are trying to make sure that you have driven *all* the water of hydration from the compound. This is called *heating to constant weight.*
8. Using your compound's final weight, calculate its water of hydration. Follow the directions on the report page and use as a model the sample calculation that we gave previously.

QUESTIONS

1. A commercial for a well-known cleaning powder says that it "shakes out white, then turns blue." What type of reaction is occurring? What compound can one use to produce this effect?

2. A student performing the water of hydration experiment can't seem to get his hydrate to constant weight. Several times he heats the compound, allows it to cool, and weighs it. But each time he does, he obtains a different weight. Can you suggest some possible causes for this dilemma?

Report on Experiment 13

Name _____

Section _____ Date _____

Instructor _____

Unknown number _____ Molecular weight of anhydrous salt _____

Quantity	How to Obtain	Data
1. Wt. of test tube	Weigh	
2. Wt. of test tube + hydrated salt	Weigh	
3. Wt. of hydrated salt	(2) − (1)	
4. Wt. of test tube + salt after first heating	Weigh	
5. Wt. of test tube + salt after second heating	Weigh	
6. Wt. of test tube + salt after third heating	Weigh	
7. Wt. of test tube + anhydrous salt	Same as (6) unless you had to heat more	
8. Wt. of anhydrous salt	(7) − (1)	
9. Wt. of water driven from the compound	(3) − (8)	
10. Moles of anhydrous salt	$\dfrac{(8)}{\text{M.W. of anhydrous salt}}$	
11. Moles of water	$\dfrac{(9)}{18 \text{ g/mole}}$	
12. Moles of water of hydration, n	$\dfrac{(11)}{(10)}$	
13. General formula for the hydrated salt	$MX \cdot nH_2O$ (put in your n to the nearest whole number)	

RESPONSES TO QUESTIONS

1.

2.

Experiment 14
The Acetic Acid Content of Vinegar

Time: About 2 hours

Materials: Commercial colorless vinegars, burettes, Erlenmeyer flasks, phenolphthalein indicator solution, standard sodium hydroxide solution (0.05 g of NaOH per mL of solution)

Safety and Chemical Hazard Guide

Please review this before performing the experiment.

1. Glasses must be worn at *all* times when in the laboratory!
2. Vinegar. Avoid splashing into eyes or on open cuts. Although vinegar presents no special hazards, more concentrated solutions of acetic acid should be handled with extreme caution. Concentrated solutions of acetic acid can cause asphyxiation, nausea, vomiting, chest pains, weakness, stupor, convulsions, and eye irritations. Avoid inhalation, ingestion, and skin contact.
3. Sodium hydroxide (1.25M). CAUTION. This chemical can cause skin burns on contact. Wash affected area immediately with plenty of water, and notify your instructor. It can cause eye, nose, and throat burns, and temporary loss of hair on the skin, upon contact.
4. Phenolphthalein indicator solution. Usually available as a one-percent solution in ethanol. Avoid ingestion, inhalation, and skin contact. Phenolphthalein has laxative properties when ingested.

INTRODUCTION

Many of the products we use around our homes contain a great variety of chemicals. Not only cleaning materials, clothing, and drugs, but also food products contain chemicals. The company that produces a certain consumer item must make sure that the proper amount of each ingredient has been added; otherwise the product may not function properly. Adding too much or too little of a chemical could produce health problems, especially if the product happens to be vitamin pills or other pharmaceuticals. Companies are concerned not only about the health factor, but about the uniformity of their products; they want you, the consumer, to know what you are getting each time you buy the products. You certainly wouldn't want to buy a brand of ketchup that had a different color and a different taste each time you bought another bottle of it.

To ensure uniformity, most companies have quality control laboratories. In these laboratories chemists and technicians check samples of each product to be sure that the product meets rigid specifications. Regardless of whether the product is perfume, toothpaste, vitamin pills, or processed food items, quality control personnel check the main chemical ingredients.

Vinegar is a household staple that you undoubtedly put on your salads or use to season many other dishes. It consists mostly of water, along with acetic acid, salt, herbs, and spices. Acetic acid is the substance in vinegar that gives it its characteristic taste and odor. It is very important that the amount of acetic acid in vinegar be kept around 4 to 5 percent by weight. The Department of Agriculture requires that the acetic acid content must be at least 4 percent, but if the acetic acid content gets too high, the vinegar won't taste good. In this experiment you will act as a quality control chemist. Your task is to check the acetic acid content of vinegar. You will do this by analyzing samples of vinegar from your neighborhood supermarket. The analysis involves performing a titration. Before you try, your instructor will demonstrate the procedure.

Some Things You Will Learn by Doing This Experiment

1. You will learn how to use a burette.

2. You will learn how to perform an acid-base titration.

3. You will learn the principle behind an acid-base titration.

4. You will learn how to use titration data to perform the calculations for determining the percent acetic acid in vinegar.

DISCUSSION

Titration analysis is one of the most important and useful types of analysis in the field of analytical chemistry. The material to be analyzed is usually in solution and is allowed to react with a reagent solution whose concentration is known. The analyst gradually adds the reagent solution to the sample from a burette. A burette is a measuring device that enables one to add the reagent solution slowly, until it just reacts with all of the constituent in the solution being analyzed. When this point—called the *equivalence point* of the titration—is reached, the analyst stops adding reagent solution. How is the equivalence point recognized? The answer lies in a substance called an *indicator,* which is usually placed in the solution being analyzed before the titration begins. The indicator does not interfere with the analysis; however, at the end of the titration, the indicator signals the equivalence point by turning a color (or changing from its initial color). From the burette, the analyst reads the volume of the reagent solution that is used to react with the constituent in the sample. By finding out the volume of reagent solution required, the analyst can determine the amount of constituent in the solution being analyzed.

The acetic acid content of vinegar is your subject for analysis by titration. In this experiment the reagent solution contains sodium hydroxide of known concentration, which reacts with the acetic acid portion of the vinegar. The analyst adds the sodium hydroxide solution slowly from a burette until all the acetic acid has been neutralized. A chemical indicator called *phenolphthalein,* which is colorless in acid solutions and pink in basic solutions, signals the endpoint of the titration. In the untitrated sample, the phenolphthalein is colorless. At the endpoint of the titration, the phenolphthalein turns pink.

One can determine the percent acetic acid in the vinegar by using the following information:

a) The volume of NaOH used to titrate the sample.

b) The concentration of NaOH solution in grams of NaOH per milliliter of NaOH solution.

c) The stoichiometry of the reaction,

$$NaOH + HC_2H_3O_2 \rightarrow H_2O + NaC_2H_3O_2$$

M.W. = 40   M.W. = 60

which tells us that 40 g of NaOH react with 60 g of $HC_2H_3O_2$. In other words, the ratio in which these two substances react is

$$\frac{NaOH}{HC_2H_3O_2} = \frac{40\ g}{60\ g} \quad \text{or} \quad \frac{4.0\ g}{6.0\ g} \quad \text{or} \quad \frac{1.0\ g}{1.5\ g}$$

Later in the experiment, you will see why this information is important.

d) The weight of your vinegar sample.

PROCEDURE

. . . .

Note: To simplify matters, we shall assume in this experiment that the density of vinegar is 1 g/mL. So when you measure the volume of your vinegar sample in milliliters, that will also be its weight in grams.

. . . .

1. Rinse two burettes, first with detergent solution and then with water. The test for a clean burette involves filling the burette with water, then letting the water run out, and seeing whether or not any drops of water adhere to the walls of the burette. (No water should adhere to the burette.)

2. Rinse one of the clean burettes with a *small* (5 mL) amount of vinegar solution. Allow 1 mL to drain out the tip, and then, while rotating the burette slowly, pour the remainder out the top. Fill the burette with vinegar. Be sure that the tip of the burette is also filled, and that there are no air bubbles.

3. Rinse the other clean burette with a *small* amount of the standard sodium hydroxide solution that you are going to use in this analysis. Pour this rinse out and then fill the burette with the standard sodium hydroxide solution (see Figure 14-1). Again, be sure that the tip is filled and that there are no air bubbles.

4. Obtain a clean 125-mL Erlenmeyer flask and allow exactly 25.0 mL of vinegar to flow into the flask from the burette.

. . . .

Note: Take all burette readings to the nearest 0.1 mL. Notice that a burette reads in a direction opposite to that of a graduated cylinder. In other words, zero is at the top and 50 is at the bottom.

. . . .

You now have a 25.0 g sample of vinegar in your flask. Record its weight on the report page. If by chance you allowed a little more or less vinegar to flow into the flask, that's okay. However, record the proper weight of your vinegar on the report page.

Figure 14-1
Burette filled with sodium hydroxide solution, clamped into position and ready to titrate vinegar solution
in Erlenmeyer flask

5. Add about 25 mL of distilled water to the sample in the flask.

. . . .

Note: This water does not affect your sample or the titration; it just serves to wash
the sides of the flask and make the endpoint easier to see.

. . . .

6. Add two or three drops of phenolphthalein indicator solution to the flask.

7. Record the initial burette reading of the NaOH burette and then titrate your
   vinegar sample with the sodium hydroxide solution. The signal for the endpoint of the titration is a change in the color of the solution from colorless to
   pink. Record the final burette reading of the NaOH and determine the number
   of milliliters of sodium hydroxide solution used.

. . . .

Note: Proceed slowly. You do not want to overshoot the endpoint. With proper
technique, one drop of NaOH solution should change the sample solution from
colorless to pink as you reach the endpoint. If you do overshoot the endpoint, you
can still salvage your sample by doing the following.

. . . .

a) Add additional vinegar to your flask from the vinegar burette until the
   phenolphthalein turns colorless. Swirl the flask to mix the contents. The
   weight of your vinegar sample has now been increased by the amount of
   vinegar you added. Be sure to note this change on the report page when calculating the weight of the vinegar sample.

b) Continue to titrate your vinegar sample with the sodium hydroxide. This
   time, try not to overshoot the endpoint. Record the final burette reading of
   the NaOH and determine the number of milliliters of NaOH solution used.

8. Perform two more titrations, using fresh samples of approximately the same size as your first sample. Record all data on the report page.

9. For each trial, calculate the percentage acetic acid in your vinegar sample. Follow the directions on the report page. Be sure you understand each step of the calculations.

QUESTIONS

1. How would your results be affected if acids other than acetic acid were present in the vinegar? Explain your answer.

2. Why is it always a good idea to rinse your burette with a little of the solution you're going to fill it with before you fill it completely?

Report on Experiment 14

Name _____

Section _____ Date _____

Instructor _____

Quantity	How to Obtain	Trial 1	2	3
1. Initial burette reading of vinegar	Read			
2. Final burette reading of vinegar	Read			
3. Weight of vinegar sample	(2) − (1)			
4. Initial burette reading of sodium hydroxide	Read			
5. Final burette reading of sodium hydroxide	Read			
6. Milliliters of sodium hydroxide solution used	(5) − (4)			
7. Concentration of NaOH solution	0.050 g/mL			
8. Grams of NaOH in the volume of the NaOH solution used for the titration	(6) × (7)			
9. Grams of acetic acid neutralized by the NaOH solution	(8) × 1.5 g*			
10. Percent acetic acid in the vinegar	$\frac{(9)}{(3)} \times 100$			

*Remember that the ratio in which these two substances react is 1 g of NaOH to 1.5 g of $HC_2H_3O_2$.

©1988 Houghton Mifflin Company

RESPONSES TO QUESTIONS

1.

2.

Experiment 15
Water Hardness: A Titration Analysis

Time: About 2 hours

Materials: Burettes, burette stands, clamps, EDTA-Mg buffer solution, Eriochrome black T indicator mixture, 0.01M EDTA solution, flasks, beakers, 25-mL pipettes, 25-mL graduated cylinders

Safety and Chemical Hazard Guide

Please review this before performing the experiment.

1. Glasses must be worn at *all* times when in the laboratory!
2. EDTA solution (0.01M). No special hazards. However, avoid ingestion and skin contact.
3. Eriochrome black T indicator solution. This substance is a dye and may stain clothes and skin. Avoid ingestion and skin contact. Wash affected area immediately with water.
4. EDTA-Mg buffer solution. CAUTION. This buffer solution contains ammonium hydroxide, which can cause skin burns on contact. Wash affected area immediately with plenty of water, and notify your instructor. It is also an eye, nose, and throat irritant. Avoid inhalation and ingestion. Inhalation of this substance may cause bronchial spasms (coughing), chest pain, and pulmonary edema. Its target organs are the eyes, lungs, and respiratory system.

INTRODUCTION

All natural waters have salts dissolved in them. The salts give the water a unique taste. However, sometimes the water contains too much salt. Salts in water that is used for drinking or washing purposes can create problems, caused mainly by the presence of calcium and magnesium salts in the water. If these salts are present in large enough concentrations, the water tastes bad, suds won't form, and one says that this water is "hard." In this experiment you will perform a test to find the hardness of a sample of water. The analysis involves performing a titration. Before you try your hand at it, your instructor will demonstrate the technique.

Some Things You Will Learn by Doing This Experiment

1. You will learn how to use a burette.
2. You will learn how to perform a test for water hardness, given the necessary reagents.

DISCUSSION

. . . .

Note: If you have already read the discussion of titration analysis in Experiment 14, skip to the second paragraph.

. . . .

Titration analysis is one of the most important and useful types of analysis in the field of analytical chemistry. The material to be analyzed is usually in solution, and is allowed to react with a reagent solution whose concentration is known. The analyst gradually adds the reagent solution to the sample from a burette. A burette is a measuring device that enables one to add the reagent solution slowly, until it just reacts with all of the constituent in the solution being analyzed. When this point—called the *equivalence point* of the titration—is reached, the analyst stops adding reagent solution. How does the analyst know when the equivalence point is at hand? The answer lies in a substance called an *indicator,* which is usually placed in the solution being analyzed before the titration begins. The indicator does not interfere with the analysis; however, at the end of the titration, the indicator signals the equivalence point by turning a color (or changing from its initial color). From the burette, the analyst reads the volume of the reagent solution that is used to react with the constituent in the sample. By finding out the volume of reagent solution required, the analyst can determine the amount of constituent in the solution being analyzed.

In this experiment you are going to determine the hardness of water by titration, using a reagent solution containing a chemical called *ethylene diamine tetraacetic acid* (EDTA for short). The EDTA reacts with the calcium and magnesium ions in the water. (It is these calcium and magnesium ions that give the water its hardness.) The analyst adds the EDTA slowly from a burette until all the calcium and magnesium ions have been reacted. This point is signaled by a chemical indicator called Eriochrome black T. In the untitrated sample, the Eriochrome black T is red. At the endpoint of the titration, the Eriochrome black T turns blue. The experimenter adds a buffer solution because the reaction between the EDTA and the calcium and magnesium ions is quantitative only in a very basic solution, with a pH of about 10.

Chemists measure water hardness in terms of ppm of $CaCO_3$. (The abbreviation ppm stands for *parts per million;* 1 ppm is the same as 1 mg/liter.) You can determine the hardness of your sample by measuring the volume of your sample and the number of milliliters of EDTA used in the titration. Commercial EDTA solution is such that 1 mL of EDTA solution equals 1 mg $CaCO_3$ hardness. The formula for calculating the hardness of the water sample is

$$\text{ppm } CaCO_3 \text{ hardness} = \frac{\text{mL of EDTA used in titration}}{\text{mL of sample}} \times 1{,}000$$

PROCEDURE

1. Pipette 25 mL of water sample into a 250-mL Erlenmeyer flask.

. . . .

Note: Your instructor will demonstrate the use of a pipette. Never pipette by mouth, always use a pipette bulb!

. . . .

2. Using a graduated cylinder, add another 25 mL of *distilled* water to the flask.

. . . .

Note: This water does not affect your sample or the titration; it just serves to wash the sides of the flask and make the endpoint easier to see.

. . . .

3. Using an eyedropper, add 20 drops (approximately 1 mL) of buffer solution to the flask.

4. Using a spatula, add a small amount of indicator (about the size of a pea) to the flask.

. . . .

Note: Depending on how your indicator powder was prepared, you might want to add slightly more or less than the amount given above. The important thing is that you add enough indicator powder to your sample so that its color is wine red.

. . . .

5. Fill a clean burette with the EDTA solution. (Directions for cleaning a burette are given in Step 1 of the procedure for Experiment 14.) Be sure that the entire burette, including the tip, is filled with the EDTA and that there are no air bubbles.

6. Record the initial burette reading. Titrate the solution in the Erlenmeyer flask with the EDTA in the burette. You will know when you reach the endpoint of the titration when you see a change in color from wine red to blue. Record your final burette reading and determine the number of milliliters of EDTA needed to titrate your sample.

. . . .

Note: Proceed with the titration slowly. You do not want to overshoot the endpoint. With proper technique, one drop of EDTA will change the sample solution from wine red to blue. If you do overshoot the endpoint, discard the sample and begin again from Step 1.

. . . .

7. Repeat the procedure with a new sample and check your results.

8. Calculate the hardness of your sample by using the formula:

$$\text{ppm } CaCO_3 \text{ hardness} = \frac{\text{mL of EDTA used in titration}}{\text{mL of sample}} \times 1{,}000$$

QUESTIONS

1. The reason one uses a buffer solution is that the pH of the sample being titrated must be about 10. Why would the analysis give incorrect results if the pH of the sample were increased to 12 by the addition of hydroxide ions? (*Hint:* Consider the ions in the water sample, which cause the hardness.)

2. What would you do in the following instance?

   Suppose that you have a water sample that you want to test for hardness. You titrate the sample, but find that it is so hard that even after you have added all the EDTA in the burette the endpoint is not reached. Aside from continuing to refill the burette time and time again (which, by the way, is not a good analytical technique), what would you do to perform the analysis?

3. People talk about the water in their community as being hard, very hard, soft, and so forth. The following table classifies water on the basis of its degree of hardness. After looking at this table, classify the sample of water that you have just analyzed.

Type of Water	Hardness Range (ppm of $CaCO_3$)
Soft	0 to 60
Moderately hard	61 to 120
Hard	121 to 180
Very hard	More than 180

Report on Experiment 15

Name _____

Section _____ Date _____

Instructor _____

Quantity	Trial	
	1	2
1. Volume of water sample		
2. Initial burette reading of EDTA solution		
3. Final burette reading of EDTA solution		
4. Milliliters of EDTA solution used to titrate samples		
5. Hardness of water sample		

CALCULATIONS

1.

2.

3.

Experiment 16
Charles's Law: A Look at One of The Gas Laws

Time: About 2 hours

Materials: 125-mL Erlenmeyer flasks, thermometers, glass tubing, 2-liter beakers, graduated cylinders, 2-hole rubber stoppers (to fit the Erlenmeyer flasks)

Safety and Chemical Hazard Guide

Please review this before performing the experiment.

1. Glasses must be worn at *all* times when in the laboratory!
2. There are no chemicals used in this experiment. However, be careful setting up the test apparatus. Use caution when inserting glass tubing through rubber stoppers.

INTRODUCTION

In the 1780s the French physicist Jacques Charles discovered the relationship between the expansion of gases and their increasing temperature. Charles found that for each 1°C increase, a gas at 0°C expands 1/273 of its volume, and, if cooled 1°C, it shrinks 1/273 of its volume. He based all these calculations on the condition that the pressure of the gas is held constant. In this experiment you will test the relationship discovered by Charles.

Some Things You Will Learn by Doing This Experiment

1. You will learn how to set up an experimental apparatus to test Charles's Law.
2. You will learn to calculate the final volume of a gas, given the initial volume, the initial temperature, and the final temperature, using the Charles's Law formula.

DISCUSSION

If we took 1 liter of a gas at 0°C and kept cooling it more and more (at constant pressure) until the gas finally got down to −273°C, its volume would shrink to 0 mL.

. . . .

Note: This is of course theoretical, since all known substances liquefy before they reach this temperature.

. . . .

This temperature, −273°C, is called absolute zero. It serves as the basis of a new temperature scale called the Kelvin scale, named after Lord Kelvin, the physicist who proposed it. A Kelvin degree has the same temperature interval as a Celsius degree. The relationship between the two temperature scales is

°K = °C + 273

If one uses the Kelvin scale, one can state Charles's Law as follows:

The volume of a given amount of gas varies directly with the temperature in degrees Kelvin, providing the pressure remains constant.

Mathematically, one can state Charles's Law this way:

$$\frac{V_i}{V_f} = \frac{T_i}{T_f}$$

where T stands for temperature in degrees Kelvin, V stands for volume, and the subscripts i and f stand for initial and final.

Charles's Law enables one to predict the amount that the volume of a gas changes with temperature without doing an actual experiment.

Example A gas occupies a volume of 1,000 mL at 27°C. Determine its volume when the temperature is increased to 127°C. (Assume that the pressure remains constant.)

Solution Organize the data and change degrees Celsius into degrees Kelvin. (In the notation that follows, a small t will mean degrees Celsius, and a capital T will mean degrees Kelvin.)

V_i = 1,000 mL (exactly) V_f = ?

$T_i = t_i + 273 = 27 + 273 = 300°K$ $T_f = t_f + 273 = 127 + 273 = 400°K$

$$\frac{V_i}{V_f} = \frac{T_i}{T_f}$$ $$\frac{1,000 \text{ mL}}{V_f} = \frac{300°K}{400°K}$$

$$V_f = \frac{(1,000 \text{ ml}) (400°\cancel{K})}{300°\cancel{K}} = 1,333 \text{ mL}$$

Thus you can see that increasing the temperature of a gas causes the gas to increase in volume.

In the experiment you are about to perform, you are going to test Charles's Law. You'll start with an Erlenmeyer flask containing air at room temperature. The air, of course, fills the entire flask and is equal to the volume of the flask. You will heat the flask. That heating will cause the air in the flask to expand; but since the walls of the flask can't expand, the air is pushed out of the open top of the flask. If you measure the amount of air that escapes from the flask, you can compute the volume the air has at its new higher temperature. This volume is simply the sum of the volume of the flask plus the volume of the air that escaped.

You can now use Charles's Law to calculate what the final volume ought to be and see whether it checks with the final volume you obtained by your experiment.

How can one measure the volume of air that escapes from the open flask? You'll do this by channeling the air into a graduated cylinder filled with water. The air displaces the water from the graduated cylinder, and you simply read the volume displaced by reading the markings on the cylinder. Further details of the experiment are explained in the following section.

PROCEDURE

1. Assemble the apparatus as shown in Figure 16-1.

2. Make sure that the J portion of the glass tubing extends about 3 cm below the bottom of the Erlenmeyer flask, as Figure 16-1 shows.

3. Obtain a 2-liter beaker and fill it about two-thirds full with water.

4. Heat the water in the beaker to about 80°C.

5. While the water in the beaker is heating, obtain a 50-mL graduated cylinder.

6. Fill the graduated cylinder to the very top with water, invert it, and lower it into the beaker whose water you are still heating. Be sure that no air bubbles get into the graduated cylinder when you invert it and place it into the beaker. (You may want to use a piece of aluminum foil or wadded-up paper towel to cover the top of the graduated cylinder during this operation.)

7. After the water in the beaker and the graduated cylinder have reached a temperature of about 80°C, stop heating.

8. Position the beaker and its contents under the Erlenmeyer flask.

9. Quickly read the temperature of the air inside the Erlenmeyer flask. This is your initial temperature, t_i. Lower the flask into the beaker. Have your partner position the J tube so that it is in the graduated cylinder (see Figure 16-2).

10. As the hot water warms the air in the Erlenmeyer flask, the air expands. You will see this expanding air pushing water out of the graduated cylinder.

Figure 16-1
Apparatus for Charles's Law

©1988 Houghton Mifflin Company

Figure 16-2
Lowering flask into beaker of hot water and positioning J tube into graduated cylinder that contains hot water

11. When no more air seems to be entering the graduated cylinder, measure the temperature of the water. You may assume that this is the same as the temperature of the hot air in the flask. This is the final temperature of the flask, t_f.

12. To measure the amount of air in the graduated cylinder, slowly lift it from the J-shaped tube. Adjust the level of the water remaining in the graduated cylinder to the level of the water in the beaker by raising or lowering the graduated cylinder until both water levels are the same. This corrects for pressure differences inside and outside the graduated cylinder. Read the amount of air in the graduated cylinder.

. . . .

Note: What is collected in the graduated cylinder is not pure air, but a mixture of air and water vapor. You will make the correction to obtain the volume of dry air when you do the calculations.

. . . .

13. Figure out the initial volume of air in the Erlenmeyer flask. Note that just because you used a 125-mL flask does not mean that it holds 125 mL. Erlenmeyer flasks are not calibrated exactly—and besides, you had a rubber stopper and thermometer in the flask. To find the initial volume of air in your Erlenmeyer flask, fill it completely with water. Place the rubber stopper (with the thermometer, but without the glass tubing) in the flask. Some water is pushed out of the flask through the open hole in the rubber stopper.

14. Remove the rubber stopper and measure the amount of water remaining in the flask by pouring it into a graduated cylinder. The volume of the water you measure is equal to the initial volume of air, V_i.

15. Enter all data on the report page. Determine the final volume of air from your experiment. This is simply the initial volume of air, V_i, plus the volume of air collected in the graduated cylinder. Then figure out what the final volume of air should be, using Charles's Law. The two values should be very close. Calculate the percent error between your experimental value and your calculated value.

. . . .

Note: If you have forgotten how to calculate percent error, see Experiment 9, "Questions," Question 1.

. . . .

QUESTIONS

1. When working with Charles's Law, the temperature must be in degrees

   _____ .

2. Charles's Law states that the temperature of a gas is proportional to its volume

   when the _____ is constant.

3. A gas has a volume of 5 liters when it is at a temperature of 227°C. What will the volume of the gas be when the temperature is increased to 727°C?

Report on Experiment 16

Name _____

Section _____ Date _____

Instructor _____

Quantity	How to Obtain	Your Value
1. Initial temperature of air in the flask, T_i	$t_i + 273$	
2. Final temperature of air in the flask, T_f	$t_f + 273$	
3. Initial volume of air in the flask, V_i	V_i	
4. Volume of gas (air + vapor) in the graduated cylinder, V_g	V_g	

We want to determine the volume of *dry* air collected in the graduated cylinder. To do this, we must remove the error introduced by the presence of water vapor. Steps 5–8 show the calculations necessary to do this.

5. Obtain the barometric pressure, P_i	P_i	
6. Obtain the pressure of water vapor at the final temperature from Table 9 in the Appendix of this laboratory manual, P_w	P_w	
7. Calculate the pressure in the graduated cylinder due to the air alone, P_a	$P_a = P_i - P_w$	
8. Use Boyle's Law to obtain the volume of the dry air in the graduated cylinder, V_a	$V_a = \dfrac{(P_a)\,(V_g)}{P_i}$	
9. Final volume of air, V_f	$V_f = V_i + V_a$	
10. Final volume of air calculated from Charles's Law	$V_f = \dfrac{(V_i)\,(T_f)}{T_i}$	
11. Percentage error		

CALCULATIONS:

Report on Experiment 16

Name _____

Section _____ Date _____

Instructor _____

RESPONSES TO QUESTIONS

1.

2.

3.

Experiment 17
Melting Points and Boiling Points of Some Organic Compounds

Time: About 2 hours

Materials: Mineral oil, benzoic acid, citric acid, naphthalene, para-dichlorobenzene (1,4-dichlorobenzene), urea, ethyl alcohol, chloroform, isopropyl alcohol, thermometers, melting point capillary tubes

Safety and Chemical Hazard Guide

Please review this before performing the experiment.

1. Glasses must be worn at *all* times when in the laboratory!
2. Mineral oil. No special hazards. However, *hot* mineral oil can cause severe burns if spilled on the skin.
3. Benzoic acid. This substance is a mild irritant to the skin, eyes, and mucous membranes. Avoid ingestion and skin contact. Wash affected area immediately with water.
4. Citric acid. No special hazards.
5. Naphthalene. Poisoning may occur by ingestion of large doses, inhalation, or skin absorption. It may also cause nausea, vomiting, and headache. Avoid ingestion, inhalation, and skin contact. Wash affected area immediately with water.
6. Para-dichlorobenzene. Its vapors may cause irritation to the skin, throat, and eyes. Prolonged exposure to high concentrations may cause weakness and dizziness. Avoid ingestion, inhalation, and skin contact. Wash affected area immediately with water.
7. Urea. No special hazards.
8. Ethyl alcohol. This chemical is intoxicating, so do not inhale. Ingestion of *large* amounts can cause nausea, vomiting, flushing, mental excitment or depression, drowsiness, impaired perception, incoordination, stupor, coma, or death.
9. Chloroform. Inhalation of *large* doses of chloroform may cause hypotension and respiratory depression, myocardial depression, and death. Avoid ingestion, inhalation, and skin contact. Wash affected area immediately with water.

10. Isopropyl alcohol. Prolonged inhalation may cause drowsiness. This substance may be mildly irritating to the eyes, nose, and throat. Avoid ingestion and skin contact. Wash affected area immediately with water.

INTRODUCTION

When it comes to identifying compounds, two of the most useful physical properties are the *melting point* and the *boiling point.*

The melting point of a substance is the temperature at which the substance changes from a solid to a liquid.

The boiling point of a substance is the temperature at which the vapor pressure of a liquid just exceeds the pressure of the atmosphere above it. At the boiling point, a liquid substance changes to a gas.

Some Things You Will Learn by Doing This Experiment

1. You will develop a technique for determining the melting points and boiling points of compounds using the methods described here.
2. You will learn to define melting point and boiling point.
3. You will learn how to perform a mixed melting point as a method of identifying an unknown substance.

DISCUSSION

PART 1: MELTING POINTS

In Part 1 of this experiment you are going to find the melting points of some organic compounds. We chose organic compounds because they tend to have lower melting points than inorganic compounds, and this makes the experiment easier and safer to perform.

Most pure compounds have melting points that are sharp, usually within a 1-degree range. If the compound is impure, the melting point is not sharp. The compound may melt anywhere within a range of 5, 10, or 20 degrees. You can therefore use melting points to check on the purity of a known compound. You can also use melting points to identify an unknown compound. You simply mix the unknown compound with the compound that you think it to be. If you're correct, the melting point of this mixture will be the same as the melting point of the unknown compound alone. It will also be a sharp melting point. If you're wrong, the melting point of the mixture will not be the same as that of the unknown compound, nor will it be sharp. There will be a melting range, instead. The beginning of this melting range will be at a temperature that is lower than the melting point of either compound in its pure state.

PART 2: BOILING POINTS

In Part 2 of this experiment you are going to measure the boiling points of some organic compounds, as well as the boiling point of water.

Figure 17-1
Obtaining the melting point of a compound

The boiling point of a pure substance is constant. However, the temperature at which a pure substance boils *can* change, because the boiling process is affected by atmospheric pressure. And as you know, atmospheric pressure can change. Boiling points of substances listed in the *Handbook of Chemistry and Physics* are taken, for the most part, at 760 torr (1-atmosphere pressure). This is called the *normal boiling point* of the substance.

PART 3: DETERMINING MELTING POINTS*

PROCEDURE

1. Obtain a *dry* 250-mL beaker and fill it halfway with mineral oil.

. . . .

Note: Make sure your beaker is a dry beaker; oil and water do not mix, and spattering will result when you heat the oil if there is water present.

. . . .

2. Obtain a melting-point capillary tube and fill it to a height of 1 centimeter with para-dichlorobenzene.

. . . .

Note: Your instructor will demonstrate how to fill a capillary tube.

. . . .

*Your instructor may have you use a hot plate in place of a gas burner in this part of the experiment.

Figure 17-2
Obtaining the boiling point of a compound

3. Attach the filled capillary tube to a thermometer with a small elastic band. Place the thermometer with the capillary tube attached into the beaker filled with mineral oil (Figure 17-1).

4. As you slowly heat the mineral oil over your gas burner, have your partner stir the mineral oil with a stirring rod so that the heat is evenly distributed. Continue to heat until the para-dichlorobenzene in the capillary tube starts to melt. Read the thermometer to the nearest degree and record the melting point (or melting range) of the para-dichlorobenzene.

. . . .

Note: It is *very* important to heat the mineral oil slowly; otherwise you will overshoot the melting point of the substance. Do not let your rate of heating exceed 3°C per minute.

. . . .

5. Repeat the procedure with each of the following substances in the order they are listed. The compounds are arranged in order of increasing melting point.
    a) Naphthalene
    b) Benzoic acid
    c) Urea
    d) Citric acid

6. Record each of your experimentally determined melting points on the report page.

7. In the *Handbook of Chemistry and Physics,* look up the melting point for each of the compounds (see the section on "Physical Constants of Organic Compounds").

8. Obtain an unknown substance from your instructor. It will be one of the five substances whose melting points you have just determined. Determine the melting point of the unknown substance and identify the substance by performing a mixed melting point.

PART 4: DETERMINING BOILING POINTS*

PROCEDURE

1. Arrange the apparatus as shown in Figure 17-2. Note that the test tube is clamped at the neck. Also note that the bottom of the thermometer is placed about 5 cm above the bottom of the test tube.
2. You will take the boiling point of several substances, using about 2 or 3 mL of sample. The amount of sample used is small for safety reasons because some of the substances are flammable.

. . . .

Caution: *Be sure to wear your safety glasses when performing this experiment.*
. . . .

3. You will begin with water as the first sample. Since water is nonflammable, practice taking its boiling point until you develop a good technique.
4. Remove the test tube from the apparatus shown in Figure 17-2 and add 2 to 3 mL of water. Put the test tube containing the water back into position.
5. Heat the water slowly until it boils gently. As the water vapor travels toward the thermometer bulb, the mercury level in the thermometer will rise. When the thermometer reading remains constant, record this as the boiling point of the liquid. (If your sample completely vaporizes before you get a constant reading, repeat the experiment.)

. . . .

Note: Violent boiling will cause the liquid to be ejected from the test tube. Practice with water until you have learned to control the rate of boiling.
. . . .

6. Using a clean, dry test tube, repeat the procedure, with each of the following liquids:
   a) Ethyl alcohol (*caution: it's flammable*)
   b) Chloroform (*caution: it's flammable*)
   c) Isopropyl alcohol (*caution: it's flammable*)
7. Record each of your experimentally determined boiling points, to the nearest degree, on the report page.
8. In the *Handbook of Chemistry and Physics,* look up the boiling point of each compound and record these on the report page.

QUESTIONS

1. We have stated that the boiling point of a substance can be affected by atmospheric pressure.
   a) Explain how the boiling point of a substance will be affected by *increasing* the atmospheric pressure.
   b) Explain how the boiling point of a substance will be affected by *decreasing* the atmospheric pressure.

*Your instructor may have you use a hot plate in this part of the experiment.

2. You were asked to determine the melting point of an unknown compound in order to identify it as one of the compounds you had previously examined. What else could you have done to make sure that your unknown was indeed the compound you believed it to be, and not just a compound that had a similar melting point?

3. When you were determining boiling points, you were told to repeat the experiment if your sample completely vaporized before you were able to obtain a constant temperature. Why?

4. In the determination of melting points, why can't you use water instead of mineral oil?

Report on Experiment 17

Name _____

Section _____ Date _____

Instructor _____

DETERMINING MELTING POINTS

Compound	Experimentally Determined Melting Point, °C	Handbook Melting Point, °C
Para-dichlorobenzene (1,4-dichlorobenzene)		
Naphthalene		
Benzoic acid		
Urea		
Citric acid		

Unknown _____ Melting point = _____ Name of compound = _____

DETERMINING BOILING POINTS

Compound	Experimentally Determined Boiling Point, °C	Handbook Boiling Point, °C
Water		
Ethyl alcohol		
Chloroform		
Isopropyl alcohol		

RESPONSES TO QUESTIONS

1.

2.

3.

4.

Experiment 18
The Personal Air Pollution Test:
The Analysis of Solids
in Cigarette Smoke

Time: About 2 hours

Materials: Cigarettes (filter tip, nonfilter tip, and "little cigars"), 250-mL suction flasks, filter paper, glass tubing, rubber tubing, 1-hole rubber stoppers

Safety and Chemical Hazard Guide

Please review this before performing the experiment.

1. Glasses must be worn at *all* times when in the laboratory!
2. Cigarettes (various brands). Perform this experiment in a well-ventilated area, and avoid inhalation of smoke.

INTRODUCTION

The average person breathes 35 pounds of air each day. We need air to provide oxygen for our blood, which uses oxygen to carry out metabolic processes essential to keep us alive. The oxygen is ultimately converted to carbon dioxide.

However, we in the United States pollute our air with more than 280 million tons of aerial garbage each year. This dirty air wastes more than 16 billion dollars in ruined crops, property, building materials, and health. Also, dirty air shortens our lives. It is a major factor in emphysema, which kills nearly 50,000 people in the United States annually, and is one of the chief causes of chronic bronchitis, which affects about 1 out of 5 men between the ages of 40 and 60.

Yet with all the dirty air we are forced to breathe, millions add to the problem by creating additional air pollution. This additional air pollution, which we call personal air pollution, is that which results from cigarette smoking. The Surgeon General of the United States has published research proving that smoking is a direct cause of lung cancer. In fact, the death rate from lung cancer is about ten times higher for cigarette smokers than for nonsmokers. Cigarette smoke irritates the bronchial tubes and renders them more susceptible to disease. Some of the solids in cigarette smoke are carcinogenic, which means they can cause cancer. This is one reason why cigarette smoking has been blamed for cancer of the larynx (voice box). There is also strong evidence linking smoking with heart disease.

A major component of cigarette smoke is carbon monoxide (CO), which robs the body of needed oxygen and reacts primarily with the hemoglobin in the blood. The affinity of hemoglobin for carbon monoxide is more than 200 times greater than it is for oxygen. This means that carboxyhemoglobin (COHb) is a more stable compound than oxyhemoglobin (O_2Hb). That is why carbon monoxide lowers the oxygen-carrying capacity of the blood.

Cigarette smokers generally have COHb levels of about 5 percent, whereas non-smokers have levels of about 0.5 percent. This means that smokers have ten times more COHb in their blood than nonsmokers. This difference is great enough to cause increased stress on the heart, impaired time discrimination, and high blood pressure.

Another major component of cigarette smoke is the solids or small particles, which the smoker inhales and which can pass through the body's defense mechanisms and destroy lung tissue. These particles range in size from a few microns (a micron is 10^{-6} meter) to submicrons, and they are responsible in part for emphysema and chronic bronchitis. Some of these particles are mineral dust (for example, lead dust), fly ash, and organic tars. When these particles enter the lung, they can irritate and destroy the alveoli (tiny sacs in the lung in which gas exchange takes place) and thus cause emphysema.

Some Things You Will Learn by Doing This Experiment

1. You will learn how to test for solids in cigarette smoke by the method described in this experiment.
2. You will learn that there is a difference in the amount of solids obtained from various types of cigarettes.

DISCUSSION

In Part 1 of this experiment you are going to determine the amount of solids in cigarette smoke. You will test:

a) a nonfilter-tip cigarette
b) a filter-tip cigarette
c) a filter-tip cigarette with its filter removed
d) a "little cigar"

In Part 2 of this experiment you are going to test for the effectiveness of the filter on a filter-tip cigarette. You will first determine the amount of solids that get through the filter when the cigarette has burned one-third of the way. You will then find out the amount of solids that get through the filter when the cigarette has burned another one-third of the way, and again after the last one-third of the way.

Remember, however, that the solids in the smoke represent only one part of the pollution from cigarettes. There are also gaseous pollutants (such as carbon monoxide) mixed with the smoke that cannot be seen and that represent additional dangers to the health of the smoker.

. . . .

Caution: Make sure that there are *no large containers* of flammable substances in the laboratory when you are performing this experiment.

. . . .

PART 1: DETERMINING SOLIDS IN CIGARETTE SMOKE

PROCEDURE

1. Obtain a 250-ml suction flask, a piece of filter paper, a 1-hole rubber stopper, some glass tubing, and a piece of rubber tubing. Assemble the equipment as shown in Figure 18-1 and attach the suction flask to a water aspirator.
2. Remove the piece of filter paper from the apparatus and weigh it.

. . . .

Note: Perform all weighings to the nearest 0.01 g.

. . . .

3. Return the filter paper to the apparatus.
4. Obtain a nonfilter-tip cigarette and weigh it.
5. Insert the nonfilter-tip cigarette into the rubber tubing attached to the testing apparatus.
6. Turn on the water aspirator and then light the cigarette.

. . . .

Note: Adjust the aspirator so that the cigarette burns slowly, over a period of 3 to 5 minutes.

. . . .

7. After the cigarette has been consumed, turn off the aspirator. (Note: You will probably have to stop burning the cigarette approximately 1 cm from its end.) Remove the filter paper from the apparatus and weigh it.
8. Weigh the portion of the cigarette that has not been consumed. (Do not weigh the ash.) Subtract this weight from the initial weight of the cigarette.
9. Record all weights on the report page.
10. Calculate the number of milligrams of solids collected per gram of cigarette consumed.
11. Repeat this procedure with the following:
    a) a filter-tip cigarette
    b) a filter-tip cigarette with its filter removed
    c) a "little cigar"

Cigarette

Rubber stopper

Rubber tubing

Glass tubing

Filter paper

Rubber tube to suction

Suction flask

Figure 18-1
Apparatus for testing solids in cigarette smoke

Figure 18-2
Dividing the tobacco portion of a filter-tip cigarette into thirds

12. Be sure to weigh each cigarette before placing it in the apparatus. Also weigh each piece of filter paper. Record all data on the report page.

PART 2: TESTING EFFECTIVENESS OF CIGARETTE FILTERS

PROCEDURE

1. Obtain a filter-tip cigarette. Using a pencil or marking pen, mark off the length of the tobacco part in thirds, as shown in Figure 18-2.
2. Place the filter-tip cigarette in the apparatus, using the same procedure you used in Part 1. Also place a fresh piece of weighed filter paper in the apparatus.
3. Turn on the aspirator and light the cigarette. Allow the cigarette to burn until one-third of the tobacco portion has been consumed. Turn off the aspirator and extinguish the cigarette with a drop of water from an eyedropper.
4. Weigh the filter paper and record the weight on the report page.
5. Replace the filter paper, turn on the aspirator, and light the cigarette. Allow it to burn until one-third more has been consumed. Turn off the aspirator and put the cigarette out.
6. Weigh the filter paper and record the weight on the report page.
7. Repeat steps 5 and 6 with the final one-third of the cigarette.

QUESTIONS

1. Which of the cigarettes in Part 1 produced the most solids? Can you offer an explanation for your results.
2. Which portion of the cigarette in Part 2 yielded the most solids? Did you expect this result? Explain.
3. Do you feel that smoking a filter-tip cigarette is much better than smoking a nonfilter-tip cigarette? Defend your answer on the basis of the experimental data you obtained.

Report on Experiment 18

Name _____

Section _____ Date _____

Instructor _____

PART 1: SOLIDS IN CIGARETTE SMOKE

Quantity	How to Obtain	Non-filter-tip Cigarette	Filter-tip Cigarette	Filter-tip Cigarette Without Filter	Little Cigar
1. Initial wt. of filter paper	Weigh				
2. Weight of filter paper after cigarette is consumed	Weigh				
3. Weight of solids collected on filter paper (in grams)	(2) − (1)				
4. Weight of solids collected on filter paper (in mg)	(3) × 1,000				
5. Initial wt. of cigarette	Weigh				
6. Weight of cigarette after burning	Weigh				
7. Weight of cigarette consumed	(5) − (6)				
8. Mg of solids collected per gram of cigarette consumed	$\frac{(3)}{(7)}$				

	Quantity	How to Obtain	Your Data
1.	Initial wt. of filter paper	Weigh	
2.	Wt. of filter paper after one-third of cigarette is consumed	Weigh	
3.	Wt. of solids collected from first portion of cigarette	(2) − (1)	
4.	Wt. of filter paper after two-thirds of cigarette is consumed	Weigh	
5.	Wt. of solids collected from second portion of cigarette	(4) − (2)	
6.	Wt. of the filter paper after third portion of cigarette is consumed	Weigh	
7.	Wt. of solids collected from third portion of cigarette	(6) − (4)	

Report on Experiment 18

Name _____

Section _____ Date _____

Instructor _____

RESPONSES TO QUESTIONS

1.

2.

3.

Experiment 19
Voltaic and Electrolytic Cells

Time: About 2 hours

Materials: Voltaic cell unit, zinc strips (for use as zinc electrodes), copper strips (for use as copper electrodes), $1M$ $ZnSO_4$ solution, $1M$ $CuSO_4$ solution, saturated NaCl solution, phenolphthalein indicator solution, copper wire, magnesium ribbon, doorbell (that operates on 1½ volts), alligator clips, $6N$ H_2SO_4 solution, voltmeter

Safety and Chemical Hazard Guide

Please review this before performing the experiment.

1. Glasses must be worn at *all* times when in the laboratory!
2. Zinc strips. No special hazards. However, inhalation of zinc fumes may result in sweet taste, throat dryness, cough, weakness, generalized aching, chills, fever, nausea, and vomiting.
3. Copper strips. No special hazards.
4. Zinc sulfate ($1M$). This substance may be irritating to the skin and mucous membranes. Avoid ingestion, inhalation, and skin contact. Wash affected area with water.
5. Copper(II) sulfate ($1M$). This substance may be irritating to the skin and mucous membranes. Avoid ingestion, inhalation, and skin contact. Wash affected area with water.
6. Sodium chloride (saturated solution). No special hazards.
7. Phenolphthalein indicator solution. This chemical is usually available as a one-percent solution in ethanol. Avoid ingestion, inhalation, and skin contact. Phenolphthalein has laxative properties when ingested.
8. Magnesium ribbon. No special hazards. However, inhalation of magnesium dust can be irritating to the respiratory system.
9. Sulfuric acid ($6N$). CAUTION. This acid can cause skin burns on contact. Wash affected area immediately with plenty of water, and notify your instructor. This substance is an eye, nose, and throat irritant, so avoid inhalation. Its target organs are the respiratory system, eyes, skin, and teeth.

Voltmeter

Porous
partition

Ag cathode

Zn anode

Ag^{1+}
NO$_3^{1-}$
AgNO$_3$
solution

Zn^{2+}
SO$_4^{2-}$
ZnSO$_4$
solution

$$Ag^{1+} + e^- \rightarrow Ag \qquad\qquad Zn \rightarrow Zn^{2+} + 2e^-$$

Figure 19-1
An example of a voltaic cell

INTRODUCTION

In 1800, while the Italian physicist Alessandro Volta was experimenting with solutions and metals that conduct electric charges, he discovered that two metals, when separated by such solutions, can be arranged so as to produce electric current. Thus Volta invented the first electrochemical battery. In his honor we call such a battery a *voltaic cell.*

A voltaic cell produces electrical energy from a chemical reaction.

In his battery Volta used the metals silver and zinc, separating them from each other by a piece of paper soaked in a salt-and-water solution. He attached wires to the silver and zinc strips. When these wires touched each other, electric current was produced (Figure 19-1). Many scientists immediately tried to see how other combinations of metals would react.

A few weeks after Volta announced his discovery, two English chemists, William Nicholson and Anthony Carlisle, performed the reverse of Volta's experiment. Volta had used a chemical reaction to produce electricity. Nicholson and Carlisle used electricity to produce a chemical reaction. They ran electric current through water, using an experimental apparatus similar to that shown in Figure 19-2. The water decomposed slowly into hydrogen gas and oxygen gas. Nicholson and Carlisle developed what today we call an *electrolytic cell,* a device that uses electricity to produce a chemical reaction.

In this experiment you are going to assemble a voltaic cell. You will then use the electricity it produces to operate an electrolytic cell. You will also assemble another voltaic cell and use the electricity it produces to operate a doorbell.

Some Things You Will Learn by Doing This Experiment

1. You will learn the definition of a voltaic cell.
2. You will learn the definition of an electrolytic cell.

Figure 19-2
An electrolytic cell

3. You will learn how to assemble a voltaic cell and measure the voltage it produces.
4. You will learn how to use a voltaic cell to operate an electrolytic cell or ring a doorbell.

DISCUSSION

One of the voltaic cells you will assemble in this experiment consists of a *zinc electrode*, immersed in a zinc sulfate solution, and a *copper electrode*, immersed in a copper sulfate solution. Current flows between the two solutions because electrons are transferred from the zinc to the copper ions in solution.

$$Zn + Cu^{2+} \rightarrow Zn^{2+} + Cu$$

$$\underset{2e^-}{\underline{\qquad\qquad}}\uparrow$$

The current produced from the voltaic cell will operate an electrolytic cell. The electrolytic cell you will operate involves the reaction between copper and water.

$$Cu + 2H_2O \rightarrow Cu^{2+} + 2OH^{1-} + H_2$$

The copper you will use will be the copper wire attached to the electrodes of the voltaic cell. The water you will use will actually be a saturated NaCl solution (because salt water is a good conductor of electricity).

In this chemical reaction, hydroxide ions are produced at one of the electrodes. We can monitor for the presence of hydroxide ions by using the chemical indicator called phenolphthalein. The phenolphthalein is colorless in acidic and neutral solutions but turns pink in basic solutions. Therefore, as hydroxide ions are produced, the phenolphthalein turns pink.

Figure 19-3
Assembling a voltaic cell unit

PART 1: ASSEMBLING A VOLTAIC CELL

PROCEDURE

1. Obtain a voltaic cell unit. The unit consists of a plastic outer portion and a removable porous cup. The inner container must be porous so that the positive and negative ions can move between the two parts of the unit and maintain electrical neutrality between both solutions.
2. Fill the outer portion of the unit about halfway with $1M$ $CuSO_4$ solution. Fill the inner container halfway with $1M$ $ZnSO_4$ solution.
3. Place a copper strip in the unit so that it is partially immersed in the $CuSO_4$ solution (Figure 19-3).

. . . .

Note: If your copper strips are not clean, you should sand their surfaces.

. . . .

4. Place a zinc strip in the unit so that it is partially immersed in the $ZnSO_4$ solution (Figure 19-3).
5. Attach a piece of copper wire to each metal electrode (Figure 19-3).
6. The cell is now operational. Determine the voltage of the cell by testing it with a voltmeter. Record the voltage of the cell on the report page.

. . . .

Note: Your instructor will demonstrate the use of a voltmeter.

. . . .

7. In Part 2 of this experiment, you will use the electricity generated by this cell to operate an electrolytic cell.

PART 2: ASSEMBLING AND OPERATING AN ELECTROLYTIC CELL

PROCEDURE

1. Obtain a 100-mL beaker and place it in about 50 mL of saturated NaCl solution.

Figure 19-4
Using a voltaic cell to operate an electrolytic cell

2. Run the copper wires attached to the electrodes of your voltaic cell into the beaker containing the salt water. Coil the ends of the copper wires to increase the surface area in contact with the NaCl solution (Figure 19-4).

3. Place a few drops of phenolphthalein indicator in the beaker. Observe any change in the color of the solution in the immediate area of the electrodes. As you operate the electrolytic cell, observe the solution for a period of 5 to 10 minutes and record your observations on the report page.

PART 3: ASSEMBLING ANOTHER VOLTAIC CELL

PROCEDURE

1. Obtain the following materials:
   a) A piece of Mg ribbon, about 25 cm long
   b) A 250-mL beaker, filled halfway with $6 N H_2SO_4$
   c) A doorbell to which is attached on one terminal a piece of coiled copper wire and, on the other terminal, a piece of copper wire with an attached alligator clip

Figure 19-5
Arrangement of materials for voltaic cell

2. Arrange the materials as shown in Figure 19-5. The magnesium ribbon is "bunched up" and is attached to the alligator clip.

3. Lower the magnesium ribbon and copper wire into the beaker containing the $6N$ H_2SO_4. The doorbell should ring.

4. The reaction that produces the electricity is

$$Mg \; + \; 2H^{1+} \rightarrow Mg^{2+} + H_2$$
$$\underset{2e^-}{\rule{1em}{0.4pt}} \uparrow$$

In this reaction electrons are transferred from the magnesium metal to the hydrogen ions of the sulfuric acid.

QUESTIONS

1. The voltaic cells you assembled in this experiment produce voltages of about 1.5 volts. How could you use voltaic cells of this nature to produce higher voltages?

2. What type of cell is an automobile battery? Explain.

3. Using library reference materials, explain how a flashlight battery operates.

Report on Experiment 19

Name _____

Section _____ Date _____

Instructor _____

PART 1: THE VOLTAIC CELL

Observations:

Cell Voltage = _____

PART 2: THE ELECTROLYTIC CELL

Observations:

PART 3: ANOTHER VOLTAIC CELL

Observations:

1.

2.

3.

Experiment 20
Radioactivity

Time: About 2 hours

Materials: Geiger counters, beta sources, alpha sources, gamma sources, lead strips of uniform thickness and size, meter sticks, calipers or centimeter rulers

Safety and Chemical Hazard Guide

Please review this before performing the experiment.

1. Glasses must be worn at *all* times when in the laboratory!
2. Alpha, beta, and gamma sources of radiation. These are extremely low level, nonlicensed sources used for instructional purposes. However, avoid all unnecessary handling of these materials and wash your hands before leaving the laboratory.

INTRODUCTION

The Geiger-Müller counter is used to detect radioactive substances. The heart of the system is the Geiger tube, which consists of a pair of electrodes surrounded by an ionizable gas. As radiation ionizes the gas, the ions that are produced travel toward the electrodes, between which there is a high voltage. These ions cause pulses of current at the electrodes, which are picked up and recorded on the counter. The wall of the Geiger tube, which acts as the cathode, is coated with a conducting material. In the center of the tube is a fine wire, the anode, which is charged to about 1,000 volts with respect to the cathode.

The space between the electrodes is filled with a gas, usually helium or argon. When a radioactive particle enters the tube, it causes the gas to ionize in a chain-reaction type of mechanism. The reaction is stopped by a mechanism called internal quenching, which occurs because a small amount of polyatomic gas is mixed with the ionizing gas. This polyatomic gas absorbs some of the energy of the electrons and positive ions in a nonreversible dissociation reaction. The overall behavior of the tube, therefore, involves two mechanisms: those causing discharge and those quenching it.

Some Things You Will Learn by Doing This Experiment

1. You will learn how to operate a Geiger counter.
2. You will learn how to determine the plateau voltage of a Geiger counter and how to take readings with this instrument.
3. You will learn something about the differences between beta rays and gamma rays.

DISCUSSION

Figure 20-1 is a diagram of a Geiger tube. When we place a radioactive sample beneath the Geiger tube and increase the high voltage slowly, the tube will reach a voltage at which it will start to count the radioactive source. This voltage is known as the *starting potential*. If we then increase the voltage a little farther, we get a tremendous rise in the counting rate. This is known as the *threshold voltage*. After we pass through the threshold range, we enter another range of voltage in which an increase in the voltage has little effect on the counting rate. This is called the *plateau voltage*. Any increase in voltage beyond this range causes a continuous discharge in the tube, which will eventually lead to its destruction. One usually sets the operating voltage of the instrument shortly after one arrives at the threshold voltage. A good operating voltage to use is the midpoint of the plateau range. This helps to preserve the life of the tube.

A problem, not only with Geiger-Müller counters, but with all radiation-detection equipment, is *background radiation*. The sources of this radiation are cosmic rays and other naturally radioactive substances in the environment. One can diminish the effect of the background radiation by shielding the tube with lead. If you don't do this, you have to figure out the radiation coming from the environment and subtract it from the radiation coming from your sample.

In this experiment you are going to determine the plateau voltage of your Geiger counter. You may then determine the background radiation in the laboratory. In addition, you will examine some of the properties of beta and gamma rays.

PART 1: DETERMINING PLATEAU VOLTAGE

PROCEDURE

1. Turn on the Geiger counter and let it warm up for a few minutes.
2. Place your alpha source in the sample compartment of the Geiger counter and position it so that it is only about 3 cm from the bottom of the Geiger tube.

Figure 20-1
Geiger tube

Figure 20-2
Graph of counts per minute versus voltage, used in determining the plateau voltage of a Geiger-Müller counter

3. If your counter has a high-voltage switch, turn it on. Increase the voltage *slowly* until the instrument starts to count the radiation being emitted by the alpha source. (This will occur at approximately 500 volts.)

4. From this point on, increase the voltage step by step, by 50 volts each time, and observe the time it takes the instrument to count to 10,000 for each voltage. *Do not* go beyond 1,100 volts without permission.

. . . .

Note: If the radioactivity of your alpha source is low, it will take too long to wait for 10,000 counts. Instead, determine the number of counts that take place in two minutes for each voltage setting.

. . . .

5. Record all the data on the report page.

6. Using the number of counts for each voltage level and the time it took to obtain them, determine the counts per minute at each voltage setting.

. . . .

Note: Some Geiger counters give direct readings in counts per minute, so there is no need to do these calculations.

. . . .

7. From your data, plot a graph of counts per minute versus voltage. The graph you obtain should be similar to the one shown in Figure 20-2. The plateau range is the place at which the graph seems to flatten out (that is, where the increase in counts per minute is the smallest for a 50-volt increase in current).

8. Choose as your operating voltage the midpoint of the plateau range. Record this voltage on the report page. Obtain your instructor's approval, then set the instrument at this operating voltage.

PART 2: DETERMINING BACKGROUND RADIATION

PROCEDURE

1. Remove all known radioactive sources from your immediate work area.

2. Using your Geiger-Müller counter, do a 5-minute count of background radiation.

3. Repeat step 2, but this time do a 10-minute count of background radiation.

4. Record all data on the report page. What are the counts per minute (cpm) for each trial? Are they similar?

5. In all subsequent measurements, subtract the background count from the count of the radioactivity of your sample. For this subtraction, use the average background count in cpm.

PART 3: PENETRATING POWER OF BETA RAYS

In this part of the experiment you are going to measure the penetrating power of beta rays in air. You will do this by moving your Geiger tube farther and farther away from your beta source. You will then make a plot of distance versus cpm. In this manner you will find out whether air can absorb beta radiation.

PROCEDURE

1. Place your beta source in the sample compartment of your Geiger counter and position it so that it is exactly 1 cm from the bottom of the Geiger tube.

2. Do a 2-minute count of the radioactivity of the sample and determine the cpm. Record the data on the report page.

3. Subtract the background count (in cpm) from the count of the radioactivity of the sample and record these data on the report page.

4. Move the Geiger tube so that it is exactly 2 cm from the sample and repeat steps 2 and 3.

. . . .

Note: You may want to move the *sample* so that it is 2 cm from the Geiger tube. In other words, you can move either the sample or the Geiger tube, so long as there is a distance of 2 cm separating them.

. . . .

5. Repeat step 4 for distances of 4, 5, 10, 15, 20, 30, 40, 50, and 100 cm.

6. Plot a graph of radioactivity (in cpm) versus distance from the Geiger tube, for your sample. Determine the penetrating power of beta rays by finding the point on your graph at which the activity of the sample reaches zero cpm. Now look at the distance this point corresponds to.

PART 4: PENETRATING POWER OF GAMMA RAYS

Gamma radiation is very different from beta radiation, since gamma rays are more energetic and have much greater penetrating ability than beta rays. Because of this greater penetrating ability, you will use lead shields of uniform thickness and size to try to stop the gamma rays from reaching the Geiger tube. You will then make a plot of the activity of your sample versus the number of lead shields. Then you will determine the penetrating power of gamma rays.

PROCEDURE

1. Place your gamma source in the sample compartment and position it so that it is exactly 5 cm from the bottom of the Geiger tube.

2. Do a 2-minute count of the radioactivity of the sample and determine the cpm. Record the data on the report page. Also correct the count by subtracting for background radiation and record this count on the report page.

3. Place a lead shield between the sample and the Geiger tube. Repeat step 2.

4. Continue to place lead shields between the sample and the Geiger tube, one at a time, until you have a total of at least 10 shields. If time permits, you may want to add still more shields.

5. Plot a graph of radioactivity (in cpm) versus number of lead shields. Determine the penetrating power of gamma rays in the same manner as you did for beta rays, except that this time you have found the number of lead shields needed to stop the radiation of the gamma source from reaching the Geiger tube.

6. Determine the thickness of a single lead shield. Multiply this thickness by the number of shields needed to stop the gamma radiation from reaching the Geiger tube. You have now determined the thickness of lead needed to stop the gamma rays.

QUESTIONS

1. Does a Geiger tube count *all* the radiation being emitted from a sample? Explain.

2. A student determines the radioactivity of her sample three times, under the same set of experimental conditions. Her results are 985 cpm, 995 cpm, 970 cpm. Explain why she obtained different results for each trial.

3. In the procedure for determining the penetrating power of beta rays, is there any other reason why the counts per minute drop off as the sample and the Geiger tube are moved farther apart? Explain.

Report on Experiment 20

Name _____

Section _____   Date _____

Instructor _____

PART 1: PLATEAU VOLTAGE

Counts	Time (minutes)	CPM	Voltage

The operating voltage is _____ volts.

PART 2: BACKGROUND RADIATION

Time (minutes)	Counts	CPM
5		
10		

Average background radioactivity = _____ cpm.

PART 3: PENETRATING POWER OF BETA RAYS

Distance between Sample and Geiger Tube (cm)	Counts	Time (minutes)	CPM	CPM (corrected for background)

Penetrating power of beta rays (from graph) = _____ cm of air.

Report on Experiment 20

Name _____

Section _____ Date _____

Instructor _____

PART 4: PENETRATING POWER OF GAMMA RAYS

Number of Lead Absorbers	Counts	Time (minutes)	CPM	CPM (corrected for background)

Penetrating power of gamma rays (from graph) = _____ lead plates

= _____ cm of lead

1.

2.

3.

Experiment 21
Organic Compounds: A Look at Some Different Kinds

Time: About 2 hours

Materials: Pentane, ethyl alcohol, diethyl ether, formaldehyde, acetone, acetic acid (5 percent), amyl acetate, methylamine, toluene, cyclopentane, cyclohexane, salicylic acid, methyl alcohol, concentrated sulfuric acid

Safety and Chemical Hazard Guide

Please review this before performing the experiment.

1. Glasses must be worn at *all* times when in the laboratory!

Note: Many chemical substances are dangerous when inhaled. Never inhale unfamiliar substances. Even when you are familiar with the substance, be careful. The proper technique for detecting an odor is to hold the test tube containing the substance about one foot away from your nose and to waft the odor toward you with your hand. *Never inhale very deeply,* and inhale only for the shortest time possible. Although we have chosen relatively safe materials for you to work with, do not breathe any of them for more than a second or two when performing the odor characteristics part of the experiment.

2. Pentane. Prolonged inhalation causes drowsiness. This substance may also act as an eye, nose, and skin irritant. Avoid prolonged exposure, and wash affected area with water.

3. Ethyl alcohol. Prolonged inhalation can cause intoxication. Avoid prolonged exposure to this substance.

4. Diethyl ether. Prolonged inhalation causes drowsiness, followed by unconsciousness. This substance is midly irritating to the skin and mucous membranes. Avoid ingestion and skin contact. Wash affected area immediately with water.

5. Formaldehyde. This substance is an eye, nose, and throat irritant. It can also irritate the skin on contact. Avoid ingestion and skin contact. Wash affected area immediately with water.

6. Acetone. Prolonged inhalation may cause headache, fatigue, excitement, and bronchial irritation, and in *large* amounts, narcosis. This substance also acts as an eye irritant. Avoid ingestion and skin contact. Wash affected area immediately with water.

7. Acetic acid (5%). No special hazards. This concentration of acetic acid is usually found in vinegar. Glacial acetic acid, however, is highly irritating to the eyes, nose, and throat and can cause severe corrosion of the mouth and gastrointestinal tract upon ingestion.

8. Amyl acetate. Prolonged inhalation may cause drowsiness. This substance is a mild eye and nose irritant. Avoid prolonged inhalation, as well as ingestion or skin contact. Wash affected area immediately with water.

9. Methylamine. Avoid prolonged inhalation. This substance is irritating to the eyes, skin, and respiratory tract. Also avoid ingestion and skin contact. Wash affected area immediately with water.

10. Toluene. Prolonged exposure may cause drowsiness and fainting. This substance may also act as an eye, nose, throat, and skin irritant. Wash affected area immediately with water.

11. Cyclopentane. Prolonged inhalation causes drowsiness. This substance may also act as an eye, nose, and skin irritant. Avoid prolonged exposure, and wash affected area with water.

12. Cyclohexane. Prolonged inhalation causes drowsiness. This substance may also act as an eye, nose, and skin irritant. Avoid prolonged exposure, and wash affected area with water.

13. Salicylic acid. No special hazards unless ingested. Absorption of large amounts can cause vomiting, abdominal pains, increased respiration, acidosis, and mental disturbances. May cause skin rashes in sensitive individuals. Avoid ingestion.

14. Methyl alcohol. Prolonged inhalation can cause drowsiness. This chemical acts as an eye irritant and causes gastric disturbances upon ingestion. Avoid ingestion and skin contact. Wash affected area immediately with water.

15. Concentrated sulfuric acid. CAUTION. This acid can cause skin burns on contact. Wash affected area immediately with plenty of water, and notify your instructor. It is also an eye, nose, and throat irritant, so avoid inhalation. Target organs are the respiratory system, eyes, skin, and teeth.

INTRODUCTION

Organic compounds—more than six million of them—make up most of the substances we see and use each day. Our food, our clothes, even our own bodies, are composed of organic compounds. Plastics, synthetic fibers, and drugs are also organic compounds.

Organic compounds fall into two basic categories: (1) those that are found in nature and (2) those that are synthesized in the laboratory. Today, chemists are working on both. While some search for the many substances in nature which have yet to be discovered, others are trying to synthesize new organic molecules in the laboratory, molecules that could turn out to be the new miracle drugs of tomorrow—or perhaps the fibers, plastics, and fuels of the future.

Some Things You Will Learn by Doing This Experiment

1. You will learn the names, formulas, and odors of some common organic compounds.

2. You will become aware of the flammability of aliphatic hydrocarbons, aromatic hydrocarbons, and nonaromatic cyclic hydrocarbons.

3. You will learn how to prepare an ester from the reaction of an alcohol and an organic acid.

DISCUSSION

With so many organic compounds in existence, we need a method of arranging them into specific classes. Fortunately, such a system does exist.

It is based on the molecular structure of the organic molecule, or more specifically, on its functional group, which is the reactive part of the organic molecule.

For example, alcohols are a class of organic compounds whose functional group is $-OH$. Ethyl alcohol has the formula CH_3CH_2-OH, and methyl alcohol has the formula CH_3-OH. Remember from your text that one can represent a class of organic compounds by using *R*-group notation, where *R* represents the carbon and hydrogen atoms attached to the functional group. Therefore we can represent alcohols as $R-OH$.

Table 21-1 lists some of the common classes of organic compounds, along with their general formulas. In this experiment you will have a chance to examine some of the classes of organic compounds and have some practical experience with them.

Table 21-1
Some Common Classes of Organic Compounds

Class	General Formula*	Example	Name Common	Name IUPAC
Alkane	C_nH_{2n+2}	CH_4	Meth*ane* (swamp gas)	Meth*ane*
Alkene	C_nH_{2n} $n = 2$ or more	$CH_2{=}CH_2$ $CH_3-CH{=}CH_2$	Ethylene Propylene	Eth*ene* Prop*ene*
Alkyne	C_nH_{2n-2} $n = 2$ or more	$H-C{\equiv}C-H$ $CH_3-C{\equiv}C-H$	Acetylene Methyl-acetylene	Eth*yne* Prop*yne*
Alcohol	$R-OH$	CH_3-OH CH_3-CH_2-OH	Methyl alcohol (wood alcohol) Ethyl alcohol (grain alcohol)	Methan*ol* Ethan*ol*

Table 21-1 (continued)

Class	General Formula*	Example	Name Common	Name IUPAC
Ether	R—O—R	CH_3—O—CH_3 CH_3—CH_2—O—CH_2—CH_3	Dimethyl ether Diethyl ether	Methoxy-methane Ethoxy-ethane
Aldehyde	$R-\underset{\underset{H}{\mid}}{C}=O$	$H-\underset{\underset{H}{\mid}}{C}=O$ $CH_3-CH_2-\underset{\underset{H}{\mid}}{C}=O$	Formalde-hyde Propion-aldehyde	Methan*al* Propan*al*
Ketone	$R-\underset{\overset{\parallel}{O}}{C}-R$	$CH_3-\underset{\overset{\parallel}{O}}{C}-CH_3$ $CH_3-CH_2-\underset{\overset{\parallel}{O}}{C}-CH_2-CH_3$	Dimethyl ketone (acetone) Diethyl ketone	Pro-pan*one* 3-pen-tan*one*
Carboxylic acid	$R-\underset{\overset{\parallel}{O}}{C}-OH$	$H-\underset{\overset{\parallel}{O}}{C}-OH$ $CH_3-\underset{\overset{\parallel}{O}}{C}-OH$	Formic acid Acetic acid	Metha-n*oic* acid Ethan*oic* acid
Ester	$R-\underset{\overset{\parallel}{O}}{C}-O-R$	$H-\underset{\overset{\parallel}{O}}{C}-O-CH_3$ $CH_3-\underset{\overset{\parallel}{O}}{C}-O-CH_2-CH_3$	Methyl formate Ethyl acetate	Methyl-methano-ate Ethyl-ethano-ate
Amine	$R-NH_2$	CH_3-NH_2 $CH_3-CH_2-NH_2$	Methyl-amine Ethyl-amine	Amino-methane Amino-ethane

*Remember that n = number of carbon atoms.

PART 1: ODOR CHARACTERISTICS OF ORGANIC COMPOUNDS

Each class of organic compounds has certain chemical and physical properties. Members of a particular class usually have similar odors and react in a similar fashion. In the first part of this experiment you are going to observe the odors of compounds from each of the different classes.

. . . .

Note: We repeat here what we have said before. *Many chemical substances are dangerous when inhaled. Never inhale unfamiliar substances.* Even when you are familiar with a substance, be careful. There is a proper technique for examining odors; your instructor will demonstrate it (or see Experiment 7). Never inhale very deeply, and inhale only for the shortest time possible. Although we have chosen relatively safe materials for you to work with, do not breathe any of them for more than a second or two. If you have an allergy or any respiratory infection, do *not* perform this part of the experiment. Instead, ask your partner to obtain the data.

. . . .

Odor can play an important part in identifying the class of an organic compound, but *the analyst must proceed with caution.*

PROCEDURE

1. Obtain your set of test tubes containing organic substances.

2. The first test tube contains pentane, an alkane. Alkanes, as well as alkenes and alkynes, are called *hydrocarbons* and are usually obtained from crude oil.

3. Briefly smell the pentane. Most hydrocarbons have an odor that you usually associate with gasoline. Describe the odor of pentane on the report page and fill in the rest of the information requested.

4. The second test tube contains ethyl alcohol, a familiar odor. Many alcohols smell like ethyl alcohol. Alcohols are important as starting materials for the synthesis of many other organic compounds. Briefly smell the ethyl alcohol and describe its odor on the report page.

5. The third test tube contains diethyl ether, generally known in hospitals as just "ether," and still used as an anesthetic. Don't breathe it for too long or you may not finish the rest of this experiment! Pour a drop of diethyl ether on your hand and notice how quickly it evaporates. Describe its odor.

6. The fourth test tube contains formaldehyde. Some aldehydes have irritating odors, while some of the long-chain ones have pleasant odors. Formaldehyde is used by biologists to preserve biological specimens. Describe its odor.

7. The fifth test tube contains acetone, a ketone. Ketones are often used as solvents in the laboratory. They are also used to make plastics. Describe the odor of acetone.

8. The sixth test tube contains acetic acid, a carboxylic acid. Organic acids are important in synthetic organic chemistry. Acetic acid is what gives vinegar its characteristic odor and taste. Describe the odor of acetic acid.

9. The seventh test tube contains amyl acetate, an ester. The odors of many fruits are due in part to their ester components. Many esters have very nice odors and are often used in perfumes. Describe the odor of amyl acetate.

10. The eighth test tube contains methylamine. Amines act as organic bases; some smell like ammonia. Be *very* careful in determining the odor of this compound. Describe its odor.

11. The two remaining test tubes contain unknowns. Try to determine the class of each unknown by its odor; then record your results.

PART 2: HYDROCARBONS AND FLAMES

In this section you are going to examine an *aliphatic hydrocarbon* and two kinds of *cyclic hydrocarbons:* aromatic and nonaromatic. Recall from your text that cyclic hydrocarbons are those in which the carbon atoms form a ring. For example, cyclohexane, which is represented by the following structure, is a nonaromatic cyclic hydrocarbon.

The aromatic hydrocarbons are similar, except that they have alternating double and single bonds between carbon atoms in the ring. Benzene can be represented by the following structure:

Benzene is the parent compound of the aromatic hydrocarbons.

When these three types of hydrocarbons burn, they do so with different results. See whether you can determine these differences in the following experiment.

. . . .

Note: *Use extreme caution during this procedure. Be sure you are wearing your safety glasses. Remove all flammable materials from your work area.* You will perform this experiment under a fume hood. Be sure to use only the amount of sample stated in the directions.

. . . .

PROCEDURE

. . . .

Note: This part of the experiment must be performed in a fume hood.

. . . .

1. Place about 1 mL of pentane in an evaporating dish. Note its odor and record it on the report page.

2. Place the evaporating dish on a ring stand and ignite the pentane, using your gas burner. Note the color of the flame and any visible products of combustion. Record the results on the report page.

3. Clean the evaporating dish.

4. Repeat steps 1, 2, and 3, using cyclopentane, then cyclohexane, and finally toluene (also known as methyl benzene). Record your results on the report page.

5. Your instructor will furnish you with an unknown. Note its odor.
6. Repeat steps 1, 2, and 3 with this material. Determine whether the liquid is an aliphatic hydrocarbon, an aromatic hydrocarbon, or a nonaromatic cylic hydrocarbon.

PART 3: A SIMPLE ORGANIC REACTION

Organic synthesis is one of the most important areas of organic chemistry. Compounds with various functional groups react together to make compounds with new functional groups. Organic synthesis has made possible many of the modern medicines, fibers, and plastics that we've all grown accustomed to using.

In this part of the experiment you are going to synthesize an organic compound. You will start with salicylic acid and methyl alcohol. The results should be surprising. You will form a compound having a functional group different from either starting material. See if you can determine the class of this compound from its odor.

PROCEDURE

1. Place about 150 mL of water in a 250-mL beaker and heat it until it boils.
2. Meanwhile, obtain a large test tube. (Large test tubes usually have a 25-mm inside diameter and are 200 mm in length.)
3. Place 1 g of salicylic acid in the test tube and then add 10 mL of methyl alcohol. Shake the test tube until the salicylic acid dissolves.
4. Carefully determine the odor of this reaction mixture.
5. Carefully add 1 mL of concentrated sulfuric acid to the reaction mixture.

. . . .

Note: Concentrated sulfuric acid is extremely corrosive. If it should come in contact with your skin, wash it off quickly with abundant amounts of water.

. . . .

6. Stir the contents of the test tube with a glass stirring rod.
7. Place the test tube in the beaker of boiling water for about two minutes. Be careful not to let the contents of your test tube boil over the sides. If necessary, turn off your gas burner to reduce the heat.
8. After heating, remove the test tube from the water bath, and smell the contents.
9. You have produced the compound methyl salicylate, commonly known as oil of wintergreen. From its odor, determine the class of this organic compound.

QUESTIONS

1. On the basis of your experience in the laboratory, do you think odor is a good and reliable method for identifying the class of an organic compound? Explain your answer.
2. Can you identify some of the combustion products from the flammability tests in Part 2 of this experiment? List each material and try to identify some of the combustion products you observed, or that you think were produced.

3. Try writing a balanced chemical equation for the reaction you performed in Part 3 of this experiment. After you do this, try writing the general equation for the synthesis of esters from alcohols and organic acids.

4. How do you think the word *ether* was derived for that class of organic compounds?

Report on Experiment 21

Name _____

Section _____ Date _____

Instructor _____

PART 1: ODOR CHARACTERISTICS OF ORGANIC COMPOUNDS

Substance	Formula	Class of Compound	Description of Odor
Pentane			
Ethyl alcohol			
Diethyl ether			
Formaldehyde			
Acetone			
Acetic acid			
Amyl acetate			
Methylamine			
Unknown no. _____			
Unknown no. _____			

PART 2: HYDROCARBONS AND FLAMES

Substance	Formula	Class of Compound	Observations
Pentane			
Cyclopentane			
Cyclohexane			
Toluene			
Unknown no. _____			

PART 3: A SIMPLE ORGANIC REACTION

1. Describe the odor of the reactants.

2. Describe the reaction.

3. Describe the odor of the product. Try to determine the class of the compound formed.

Report on Experiment 21

Name _____

Section _____ Date _____

Instructor _____

RESPONSES TO QUESTIONS

1.

2.

3.

4.

Experiment 22
Carbohydrates, Lipids, and Proteins

Time: About 2 hours

Materials: Four test tubes, 250-mL beaker, stirring rods, brown paper, butter, distilled water, 2% starch solution, iodine solution, Benedict's solution, glycerol, cottonseed oil, olive oil, potassium bisulfite, 1% biuret, 2% albumin, milk, white bread, 10% NaOH, 0.5% $CuSO_4$, diabetic urine, 2% glucose

. . . .

Safety and Chemical Hazard Guide

Please review this before performing the experiment.

1. Glasses must be worn at *all* times when in the laboratory!
2. Starch solution (2%). No special hazards.
3. Iodine solution. This material stains the skin. It acts as a poison if ingested. It is also an eye and skin irritant. Avoid ingestion and skin contact. If skin contact occurs, wash affected area with soap and water as soon as possible.
4. Benedict's solution. CAUTION. This is an alkaline substance that also contains copper(II) citrate. It may cause skin irritations. Avoid ingestion and skin contact. If contact occurs, wash affected area immediately.
5. Glycerol. No special hazards.
6. Cottonseed oil. No special hazards.
7. Olive oil. No special hazards.
8. Potassium bisulfite. Avoid ingestion, inhalation, and skin contact. Concentrated solutions are irritating to the skin and mucous membranes. Wash affected area with water immediately.
9. Biuret solution (1%). This substance is an irritant. Avoid ingestion, inhalation, and skin contact. Wash affected area immediately with water. Ingestion may produce strong irritation of the urinary tract.
10. Albumin (2%). No special hazards.
11. Sodium hydroxide (10%). CAUTION. This chemical can cause skin burns on contact. Wash affected area immediately with plenty of water, and notify your instructor. It can also cause eye, nose, and throat burns, and temporary loss of hair on the skin upon contact.

12. Copper(II) sulfate solution (0.5%). This chemical may be irritating to the skin. Avoid ingestion and skin contact. Wash affected area with water.

13. Glucose (2%). No special hazards.

INTRODUCTION

Carbohydrates, lipids, and proteins are organic compounds found in the foods we eat. These are large molecules that are broken down during digestion to form smaller molecules. The small molecules enter our cells and take part in various body processes. In this laboratory exercise, we'll perform various tests to detect the presence of carbohydrates, lipids, and proteins.

Some Things You Will Learn by Doing This Experiment

1. You will learn how to test for the presence of glucose by using the Benedict's test.
2. You will learn how to test for the presence of starch by using an iodine solution.
3. You will learn how to test for the presence of fats or oils by using the acrolein test.
4. You will learn how to test for the presence of fats or oils by using brown paper.
5. You will learn how to test for proteins by using the biuret test.

DISCUSSION

PART 1: TESTING FOR CARBOHYDRATES

Carbohydrate molecules are organic compounds that are composed of the elements carbon, hydrogen, and oxygen. Carbohydrates are one of the types of molecules that supply us with the energy to enable our bodies to function.

In this experiment we will learn to test for two carbohydrates, glucose and starch. First we'll treat several substances with Benedict's reagent. The Benedict's reagent reacts with small carbohydrate molecules, such as glucose, that have a free aldehyde or ketone group in solution and can act as reducing agents. Benedict's reagent is an alkaline solution of copper(II) sulfate that contains the Cu^{2+} ion and is blue. We'll heat the mixture. If a carbohydrate, such as glucose (which contains a free aldehyde group), is present, the Cu^{2+} ions are reduced to Cu^{1+} ions. Brick-red copper(I) oxide solid, Cu_2O, forms. The solution turns from green to orange, depending on how much glucose is present. For glucose, the reaction is as follows:

Our second experiment will be a test for starch, using iodine. When we place iodine on a starch-containing substance, we note a blue-black color.

PROCEDURE FOR THE BENEDICT'S TEST

. . . .

Note: Be sure that your safety glasses are on!

. . . .

1. Prepare a boiling-water bath by filling a 250-mL beaker with water and heating the water to the boiling point.
2. Clean and dry four test tubes and label them as shown below.

Tube no.	Solution
1	Distilled water (as control)
2	2% glucose
3	Diabetic urine
4	Your own urine

3. Add 5 mL of Benedict's solution to each test tube.
4. Add 1 mL of the solutions indicated on the label of each test tube.
5. Stir the contents of each test tube thoroughly, using a clean stirring rod for each.
6. Place all the tubes in the boiling-water bath for 10 minutes.
7. At one-minute time intervals compare the colors of the solutions with the color of the control tube (tube number 1).
8. Record your observations on the report page.

PROCEDURE FOR THE IODINE TEST

1. Clean and dry four test tubes and label them as shown below.

Tube no.	Substance
1	Distilled water (as control)
2	2% starch solution
3	White bread
4	Hard-boiled egg white

2. Add 1 mL of distilled water to tube number 1. Add 1 mL of 2% starch solution to tube number 2. Add a small piece of white bread to tube number 3. Add a small piece of hard-boiled egg white to tube number 4.

3. Place 5 drops of iodine solution in each test tube.

4. Record your results on the report page.

PART 2: TESTING FOR LIPIDS

Lipids are an important group of biochemical molecules that include fats and other compounds that have physical properties similar to those of fats. In this part of the experiment we'll learn to test for fats using two tests, the acrolein test and the brown-paper test.

All fats are esters of fatty acids and the alcohol glycerol. Fats can be broken down to fatty acids and glycerol by hydrolysis (reaction with water). Glycerol, which is an alcohol containing three −OH groups, is commonly called glycerin. When reacted with dehydrating agents or at high temperatures, glycerol forms *acrolein,* a substance that can be detected by its pungent color. If you've ever smelled burnt fat or oil, you've smelled the odor of acrolein. The reaction is as follows:

$$
\begin{array}{ccc}
\begin{array}{l} CH_2\text{−OH} \\ | \\ CH\text{−OH} \\ | \\ CH_2\text{−OH} \end{array} & \xrightarrow{\text{Heat}} & \begin{array}{l} CH_2 \\ \| \\ CH \\ | \\ C{=}O \\ | \\ H \end{array} \quad +\ 2H_2O \\[2em]
\text{Glycerol} & & \text{Acrolein}
\end{array}
$$

Fats can also be detected by a simpler, less sophisticated method. When dropped on a piece of brown paper, fats leave a greasy spot that will not dry.

PROCEDURE FOR THE ACROLEIN TEST

. . . .

Note: Are your safety glasses on?

. . . .

1. Obtain four clean, dry test tubes and label them as shown below.

Tube no.	Substance
1	Water (as control)
2	Glycerol
3	Cottonseed oil
4	Olive oil

2. Fill the rounded portion at the bottom of each test tube with potassium bisulfite.

3. Add three drops of the substance indicated on the label to each test tube.

4. Heat test tube number 1 vigorously for about 30 seconds. Very carefully, holding the tube one foot away from your nose, waft the odor coming from the test tube toward you with your hand. Note the odor.

5. Repeat steps 4 and 5 with the remaining test tubes. Record your results on the report page.

PROCEDURE FOR THE BROWN-PAPER TEST

1. Obtain a piece of brown paper and draw lines on the paper, separating it into four sections.

2. Place two drops of water on section 1.

3. Place two drops of olive oil on section 2.

4. Place two drops of cottonseed oil on section 3.

5. Place a small chunk of butter on section 4.

6. Record your results on the report page.

PART 3: TESTING FOR PROTEINS

Proteins are very large molecules that are made of carbon, oxygen, hydrogen, nitrogen, and sometimes sulfur, iodine, iron, or phosphorus. Molecules called *amino acids* link together in special arrangements to form the large protein molecules. The bonds that hold amino acids together to form proteins are called *peptide bonds*. The *biuret test* is used to detect the presence of two or more peptide bonds. We add the test solution to a small amount of copper(II) sulfate and sodium hydroxide. The solution turns purple if a dipeptide (two amino acids bonded together) or a larger peptide is present. The color produced is similar to the color that results when biuret solution is tested.

PROCEDURE FOR BIURET TEST

1. Obtain four clean, dry test tubes and label them as shown below.

Tube no.	Solution
1	Water (as control)
2	1% biuret
3	2% albumin
4	Milk

2. Add to each test tube 2.5 mL of the solution indicated on the label.

3. Add 1.0 ml of 10% NaOH to test tube number 1 and mix well.

4. Add one drop of 0.5% $CuSO_4$ to test tube number 1 and mix well. A purple color indicates the presence of a protein.

5. Repeat steps 2, 3, and 4 on the remaining test tubes. Record your results on the report page.

QUESTIONS

1. Which of the solutions tested with Benedict's reagent showed a positive result for the presence of glucose? Explain why the results for diabetic urine and your own urine were different.
2. Predict the results you would obtain if you used the iodine test on the following substances: (a) spaghetti, (b) apple juice, (c) paper.
3. Explain the principle behind the operation of the acrolein test.
4. According to the biuret test, which of the solutions tested contain peptide linkages?

Report on Experiment 22

Name _____

Section _____ Date _____

Instructor _____

PART 1: TESTING FOR CARBOHYDRATES

BENEDICT'S TEST

Tube no.	Solution	Color of Solution
1	Distilled water (as control)	
2	2% glucose	
3	Diabetic urine	
4	Your own urine	

IODINE TEST

Tube no.	Substance	Color of Solution
1	Distilled water (as control)	
2	2% starch	
3	White bread	
4	Hard-boiled egg white	

PART 2: TESTING FOR LIPIDS

ACROLEIN TEST

Tube no.	Substance	Odor
1	Distilled water (as control)	
2	Glycerol	
3	Cottonseed oil	
4	Olive oil	

BROWN-PAPER TEST

Section no.	Observation
1	
2	
3	
4	

PART 3: TESTING FOR PROTEINS

BIURET TEST

Tube no.	Solution	Color
1	Distilled water (as control)	
2	1% biuret	
3	2% albumin	
4	Milk	

Report on Experiment 22

Name _____

Section _____ Date _____

Instructor _____

RESPONSES TO QUESTIONS

 1.

 2.

 3.

 4.

Experiment 23
Determination of the Amount of Iron in Water

Time: About 2 hours

Materials: Colorimeter, standard iron solution, unknown iron solutions, FerroVer® Powder Pillows, 100-mL beakers, 25-mL graduated cylinders

Safety and Chemical Hazard Guide

Please review this before performing the experiment.

1. Glasses must be worn at *all* times when in the laboratory!
2. Iron(II) nitrate solution (10 mg/L). No special hazards.
3. FerroVer® Powder Pillows. These powder pillows contain 1,10-phenanthroline, and may be a skin irritant. Avoid ingestion and skin contact. If skin contact occurs, wash affected area with water.

INTRODUCTION

Iron can create serious problems in public water supplies. Along with manganese, iron can interfere with laundering operations and can impart stains to plumbing fixtures. Iron and manganese can also cause problems in the water-distribution system by supporting the growth of bacteria. Iron also imparts a taste to water, a taste that is detectable at very low concentrations. Because of this, the U.S. Public Health Service recommends that public water supplies should not contain more than 0.3 mg/liter of iron or 0.05 mg/liter of manganese. In this experiment we'll determine the amount of iron in various water samples by using a colorimetric procedure.

Some Things You Will Learn by Doing This Experiment

1. You will learn how to operate a colorimeter.
2. You will learn the principle behind the colorimetric determination of iron.

®FerroVer is a registered trademark of the Hach Chemical Company.

3. You will learn to plot a graph of absorbance versus concentration, and you will learn how to use this graph to find the concentration of an unknown sample when you know its absorbance.

DISCUSSION

To determine the amount of iron in a sample of water, you'll use the colorimetric method. The principle behind this method is to find a chemical agent or agents that reacts specifically with iron ions, so that a colored solution results. The intensity of the color depends on the amount of iron ions in the sample. Such chemical agents do exist for iron ions. A substance called 1,10-phenanthroline reacts with iron(II) ions to produce an orange-red complex.

1,10-phenanthroline   Orange-red complex

But what about iron that exists in water as iron(III) ions? It is possible to reduce iron(III) to iron(II) by adding the reducing agent hydroxylamine to the water sample.

$$4Fe^{3+} + 2NH_2OH \rightarrow 4Fe^{2+} + N_2O + H_2O + 4H^{1+}$$

Hydroxylamine

In this manner, we can determine the total amount of iron in a water sample.

The colorimetric procedure involves making up a series of known iron solutions and treating them with 1,10-phenanthroline. At the same time, we also treat the water sample (whose iron content is to be determined) with the 1,10-phenanthroline. After the color develops in the solutions, we compare the color of the water sample with the colors of the known iron solutions.

A person can, of course, compare colors by visual observation, but this method is not always reliable, especially if the intensity of the colors is very similar.

A much better method is to use a device called a colorimeter, an instrument that measures the intensity of the color electronically.

The colorimeter does this by measuring the absorbance of a specific wavelength of light as it passes through the colored sample.

Each of the known iron solutions absorbs a different amount of light. The more iron in the solution, the more light it absorbs. The observer reads the absorbance of each sample on the colorimeter meter, and then draws a graph, plotting the concentration of each *known* iron solution against its absorbance. The procedure should give a straight-line graph in the range of zero to three milligrams per liter of iron. This graph enables one to determine the iron content of the water sample by simply finding the absorbance of the sample on the graph and seeing what concentration of iron this absorbance corresponds to.

In our procedure we'll be using FerroVer® powder pillows. These powder pillows contain a premeasured amount of 1,10-phenanthroline plus the reducing agent needed to change iron(III) to iron(II).

1—Pilot light   5—Percent-transmission knob
2—Meter   6—Zero-control knob
3—Wavelength control   7—Sample holder
4—Wavelength scale

Figure 23-1
The Spectronic® 20 spectrophotometer

PART 1: CALIBRATION OF THE COLORIMETER

PROCEDURE

(The procedure here was written for use with the Bausch and Lomb Spectronic®
20 spectrophotometer shown in Figure 23-1. If you are using another type of
colorimeter, some minor changes in the procedure might be needed.)

1. Obtain your colorimeter and colorimeter test tubes.

2. Turn on the instrument and adjust the wavelength-control knob to 510
   nanometers. Allow the instrument to warm up for 10 minutes.

3. With the sample compartment empty, adjust the zero-control knob (the left-
   hand knob) so that the meter pointer reads infinite absorbance. (This is the
   same as zero-percent transmittance.) Close the cover of the sample holder when
   you are adjusting the meter.

4. Fill one colorimeter test tube with distilled water and place it in the sample
   compartment of the instrument. Close the cover. The meter pointer should
   move to the right.

5. Using the percent-transmittance knob (the right-hand knob), adjust the
   meter pointer to read zero absorbance. (This is the same as 100-percent
   transmittance.)

6. The instrument is now calibrated for use. You may wish to repeat the above
   procedure to be sure that the instrument is stable.

7. If you wish to learn more about the operation of the colorimeter, ask your
   instructor for a copy of the instrument manual.

PART 2: DETERMINATION OF IRON

PROCEDURE

1. Obtain the *standard iron solution* whose concentration is 10 mg/liter. Fill a clean buret with this standard iron solution. (*Note:* Be sure that you've rinsed the buret with 2 or 3 mL of this iron solution before completely filling it.)

2. Obtain four 100-mL volumetric flasks. Label these flasks as follows: 1.0 mg/liter, 1.5 mg/liter, 2.0 mg/liter, and 2.5 mg/liter.

3. Into each flask, buret the proper amount of *standard iron solution* in order to prepare the concentrations written on each flask. Proceed as shown below.

Flask no.	Concentration	Milliliters of iron solution to put into flask
1	1.0 mg/liter	10 mL
2	1.5 mg/liter	15 mL
3	2.0 mg/liter	20 mL
4	2.5 mg/liter	25 mL

4. Fill each flask with sufficient water to bring the total volume to 100 mL. The 100-mL mark is noted on the neck of the volumetric flask.

   These four flasks now contain iron solutions whose concentrations are 1.0 mg/liter, 1.5 mg/liter, 2.0 mg/liter, and 2.5 mg/liter.

5. Using a 25-mL graduated cylinder, measure 25 mL of each standard and place it into a 100-mL beaker. (*Note:* Label these beakers so that you know which beaker contains a particular iron solution.)

6. Also place 25 mL of each water sample (the unknown) into a 100-mL beaker. (Your instructor may have you analyze one or more unknown water samples.)

7. Add the contents of one FerroVer® powder pillow to each beaker, and stir each solution. If iron is present, an orange-red color develops.

8. Let each solution stand for two minutes after adding the powder pillow. Then, within ten minutes, determine the absorbance of each solution.

9. To determine the absorbance of each solution, simply fill a colorimeter test tube half full with each solution and place the tube in the colorimeter. Close the cover of the sample holder when you are taking a reading.

10. Construct a graph (use page 200 of this manual). Plot absorbance versus concentration for each of the *known* iron samples. From this graph and the absorbance of your unknown sample(s), determine the concentration of iron in your unknown sample(s).

QUESTIONS

1. In what form must iron exist in order to be measured by the phenanthroline method?

2. List some of the problems caused by high concentrations of iron in potable water.

3. What could you do with an unknown water sample that you suspected has an iron concentration of about 10 mg/liter? In other words, what would you have to do to determine the iron concentration of this sample, using the colorimetric method described in this experiment?

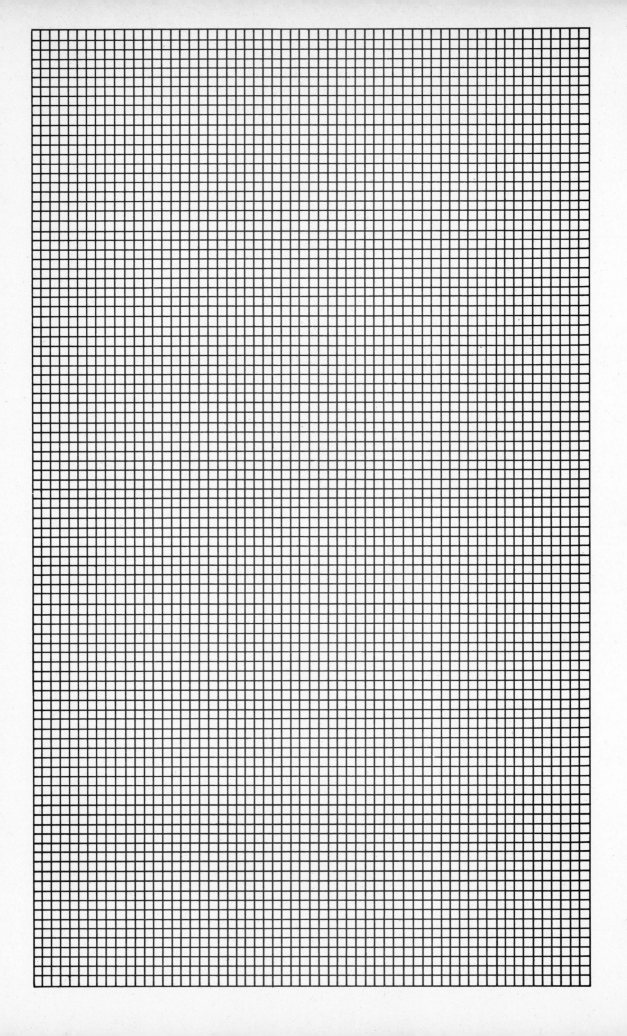

. . . .

Report on Experiment 23

Name _____

Section _____ Date _____

Instructor _____

Concentration of Known Iron Solutions (in mg/liter)	Absorbance of Known Iron Solutions
1.0 mg/liter	
1.5 mg/liter	
2.0 mg/liter	
2.5 mg/liter	

Unknown iron sample:
Sample No. _____

Absorbance of unknown iron sample: _____

Concentration of unknown iron sample from graph (mg/liter): _____

1.

2.

3.

Experiment 24
Determination of the Dissolved-Oxygen Content of Water

Time: About 2 hours

Materials: 300-mL BOD bottles (Erlenmeyer flasks that are 300 mL or larger may be used as alternatives if BOD bottles are not available), graduated cylinders (100-mL and 10-mL sizes), burets, manganese sulfate solution, alkali-iodide-azide reagent, concentrated sulfuric acid, 0.025 N sodium thiosulfate, starch solution, 500-mL Erlenmeyer flasks, plunger pipets (2) set to deliver 2 mL of solution, water samples

Safety and Chemical Hazard Guide

Please review this before performing the experiment.

1. Glasses must be worn at *all* times when in the laboratory!
2. Manganese sulfate solution. This substance may cause eye, nose, and throat irritation. Avoid ingestion, inhalation, and skin contact. Wash affected area with water.
3. Alkali-iodide-azide reagent. This solution contains concentrated sodium hydroxide. It is extremely alkaline and can cause skin burns on contact. Wash affected area immediately with plenty of water, and notify your instructor. It can also cause eye, nose, and throat burns, and temporary loss of hair on the skin upon contact.
4. Sulfuric acid (concentrated). CAUTION. This chemical can cause skin burns on contact. Wash affected area immediately with plenty of water, and notify your instructor. It is also an eye, nose, and throat irritant, so avoid inhalation. Its target organs are the respiratory system, eyes, skin, and teeth.
5. Sodium thiosulfate (0.025 N). This chemical can cause eye, nose, and throat irritations, so avoid ingestion, inhalation, and skin contact. Wash affected area with water.
6. Starch solution. No special hazards.

INTRODUCTION

If a lake or a river is to maintain a healthy condition, the amount of free dissolved oxygen (O_2) in it is extremely important. Fish and other organisms that live in water must have dissolved oxygen to survive. The amount of free oxygen that can dissolve in a sample of water is dependent on the water's temperature. The solubility of atmospheric oxygen in fresh waters ranges from 14.6 mg/liter at 0°C to about 7 mg/liter at 35°C, under one atmosphere of pressure. You can see by these data why certain bodies of water can become oxygen deficient in warm weather and produce critical situations for fish and other organisms that depend on this oxygen.

The limited amount of oxygen that can dissolve in water is also the reason why most bodies of water have limited capacities for self-purification. In other words, if we were to allow the waste products of our society to continuously flow into a lake or river, the water in that lake or river would quickly become depleted of oxygen. This would occur because these waste products act as food for microorganisms. The microorganisms would grow and multiply rapidly, consuming oxygen in the process. It wouldn't take long for the microorganism population to exceed the oxygen supply in the water. In other words, a point would be reached at which the supply of dissolved oxygen would not meet the demand. The microorganisms that require dissolved oxygen to survive would die, and so would many fish. With these microorganisms dead, the purification process of the lake or river would cease. Therefore it is important to monitor the amount of dissolved oxygen in a body of water. We shall perform such an analysis in this experiment.

Some Things You Will Learn by Doing This Experiment

1. You will learn the importance of dissolved oxygen in water.
2. You will learn how to test for the amount of dissolved oxygen in a sample of water.
3. You will learn how to perform an oxidation-reduction titration.

DISCUSSION

In this experiment we'll determine the amount of dissolved oxygen in a sample of water by titration. The procedure involves allowing an Mn^{2+} solution to react with the dissolved oxygen. The Mn^{2+} is converted to MnO_2. The amount of MnO_2 that forms depends on the amount of dissolved oxygen in the sample. The MnO_2 then reacts with a solution containing I^{1-}. The I^{1-} is oxidized to I_2. The I_2 is then titrated with a $0.025N$ sodium thiosulfate solution. The amount of sodium thiosulfate it takes to react with all the I_2 can be related back to the amount of dissolved oxygen in the sample of water. In fact, if we choose an appropriate size of sample, the amount of sodium thiosulfate used in the titration will be numerically equal to the milligrams/liter of dissolved oxygen in the sample. In other words, the milliliters of sodium thiosulfate used will be equal to the milligrams/liter of dissolved oxygen in the sample.

PROCEDURE

1. Obtain a BOD bottle and fill it to the brim with the sample. The BOD bottle requires 300 mL of sample to fill it. (If you're not using such a bottle, your instructor will supply you with a 300-mL Erlenmeyer flask or some other container.)

Try not to aerate the sample as you place it in the bottle. After filling the bottle to the brim, place the glass stopper in the bottle. Pour off the excess water.

2. Add 2 mL of manganese sulfate solution to the sample, followed by 2 mL of alkali-iodide-azide reagent. The ideal way to do this is to use a plunger pipet, calibrated to 2 mL. Make sure the tip of the pipet is well below the surface of the water sample.

. . . .

Note: If plunger pipets are not available, your instructor will have an alternative way for you to add these reagents. The important thing to remember is not to aerate the sample as you add these reagents.

. . . .

3. After addition of the two reagents, stopper the bottle carefully and invert it several times. This allows the reagents to mix with the sample. You will see the appearance of a manganese dioxide floc.

4. Allow the manganese dioxide floc to settle about halfway down the bottle. At that point, remove the stopper and add 2 mL of concentrated sulfuric acid.

. . . .

Caution: Sulfuric acid in concentrated form is extremely dangerous. Be very careful. If you spill any H_2SO_4 on your hands, wash them immediately with plenty of soap and water.

. . . .

The sulfuric acid will run down the sides of the bottle. Stopper the bottle and invert it several times so that the sulfuric acid mixes with the entire contents of the bottle.

5. At this point the contents of the bottle should clear and have a yellow or straw color. This is due to the presence of iodine. To perform the titration, we want to remove the equivalent of 200 mL of original sample. Allowing for the addition of reagents, this means that we must measure out 203 mL of sample from the bottle. Use your 100-mL graduated cylinder and your 10-mL graduated cylinder to perform this task. Pour the 203 mL of sample into a 500-mL Erlenmeyer flask.

6. Fill a buret with 0.025 N sodium thiosulfate solution. Be sure that there are no air bubbles in the tip of the buret.

7. Titrate the sample in the flask with the sodium thiosulfate until the sample turns a pale yellow color. At this point add about 2 mL of starch solution. The contents of the flask will turn blue-black as the starch complexes with the remaining iodine. Continue titrating with sodium thiosulfate until the blue color disappears.

8. Record the number of milliliters of sodium thiosulfate used to titrate the sample. This will be the same as the milligrams/liter of dissolved oxygen in the sample.

9. Repeat this procedure two or more times with the same water sample. Record you data on the report page.

. . . .

Note: Your instructor may have you perform the titration on other water samples.

. . . .

©1988 Houghton Mifflin Company

QUESTIONS

1. Explain why it is important to have an adequate supply of dissolved oxygen in a river or stream.

2. What follows is an abbreviated summary of the reactions for this titration. For each *reactant*, identify whether it is oxidized or reduced.

$$Mn^{2+} + O_2 \rightarrow MnO_2$$

$$MnO_2 + I^{1-} \rightarrow Mn^{2+} + I_2$$

$$I_2 + S_2O_3{}^{2-} \rightarrow I^{1-} + S_4O_6{}^{2-}$$

. . . .

Note: The above equations are not balanced. However, it is not necessary to balance them to answer the question.

. . . .

Report on Experiment 24

Name _____

Section _____ Date _____

Instructor _____

Sample No.	Amount of Sodium Thiosulfate Used in Titration (mL)	Dissolved-Oxygen Content of Water Sample (mg/liter)

RESPONSES TO QUESTIONS

1.

2.

Experiment 25
Chemical Kinetics: Rates of Reaction

Time: About 2 hours

Materials: 0.1M sodium oxalate, 0.1M potassium permanganate, 6M sulfuric acid, 0.2M potassium iodate (labeled solution A), 0.1M acidified sodium sulfite solution with starch (labeled solution B), deionized or distilled water, 400-mL beakers, graduated cylinders (10-mL, 25-mL, and 100-mL) or pipets (various sizes), thermometers, 0.1M copper(II) sulfate

Safety and Chemical Hazard Guide

Please review this before performing the experiment.

1. Glasses must be worn at *all* times when in the laboratory!

2. Sodium oxalate (0.1M). Ingestion of concentrated solutions of this substance may cause severe gastrointestinal irritations, as well as central nervous system and cardiac depression. Dilute solutions produce little gastrointestinal distress, but may cause weakness and muscular twitchings. Avoid ingestion and inhalation of this material.

3. Potassium permanganate (0.1M). Dilute solutions of this substance are mildly irritating, and high concentrations are caustic. Avoid ingestion, inhalation, and skin contact. Wash affected area immediately with plenty of water, and notify your instructor.

4. Sulfuric acid (6M). CAUTION. This acid can cause skin burns on contact. Wash affected area immediately with plenty of water, and notify your instructor. It is an eye, nose, and throat irritant, so avoid inhalation. Its target organs are the respiratory system, eyes, skin, and teeth.

5. Potassium iodate (0.2M). This substance may be mildly irritating to the skin. Avoid ingestion and skin contact. Wash affected area with water.

6. Sodium sulfite (0.1M). Solutions of this chemical are alkaline, and may be irritating to the skin. Avoid ingestion and skin contact. Wash affected area immediately with water.

7. Copper(II) sulfate (0.1M). Dilute solutions of this material may be irritating to the skin. Avoid ingestion and skin contact. Wash affected area immediately with water.

INTRODUCTION

Chemical kinetics is the study of the rate or speed at which a chemical reaction occurs. Some reactions take place very rapidly, whereas others can be quite slow. Every chemical reaction has its own reaction rate. Chemical kinetics also looks at the steps that atoms or molecules composing the reactants take in order to form the products. The study of the routes or steps that the reactants take to form products is called the reaction mechanism. In this experiment we are going to take a look at the rates of two different reactions.

Four factors influence the reaction rate: (1) the *nature or type of reactants involved*, (2) the *concentration*, (3) the *temperature*, and (4) the *presence of catalysts*. In order for a chemical reaction to occur, there must be a collision between reactants. Collisions between molecules and atoms may be likened to the conditions on a crowded skating rink. When many skaters are present on the rink, collisions often occur. When just a few skaters are present, collisions are less likely to occur. In other words, the rate at which the reaction occurs depends on the *concentration* or number of atoms or molecules present. When a reaction first begins, the rate of reaction is the fastest because the greatest number of reactant particles are present. As the reaction proceeds, the number of reactant particles available for collision decreases, so the reaction rate decreases. In other words, the rate of the forward reaction slows down. What about reactions that are reversible? At first the rate of the reverse reaction is zero. As the number of atoms and molecules of product increases, there are more particles available for collision and the rate of the reverse reaction increases.

What can you do to speed up a reaction if it is proceeding at an excessively slow rate? You can add an excess of a reactant, so that it becomes easier for collisions to occur.

Temperature affects reaction rate, because as molecules and atoms are heated they move faster, making collisions occur more often. *A rise in temperature of ten degrees approximately doubles the reaction rate.* Looking back at our skating-rink analogy, we can reason that when skaters are moving about on the rink at greater speeds, they are more likely to collide with one another.

Finally, the presence of a catalyst affects reaction rate. A *catalyst* is a substance that changes the speed of the reaction but remains itself unchanged at the end of the reaction. In a skating rink with a highly polished floor, skaters are more likely to collide. The slippery floor takes no part in the reaction, but it does affect the reaction rate.

Some Things You Will Learn by Doing This Experiment

1. You will learn that there are large differences in the rates of various reactions.
2. You will see how changes in temperature of reactants, concentration of reactants, and the presence of a catalyst affects reaction rate.

DISCUSSION

Factors that influence reaction rates may be found experimentally for each reaction. In this experiment we will study three of these variables: concentration of reactants, temperature, and the effect of a catalyst.

One of the systems we will study involves the reaction between excess *iodate* ion and *sulfite* ion to form *iodide* ion and *sulfate* ion.

$$IO_3^{1-} + 3SO_3^{2-} \rightarrow I^{1-} + 3SO_4^{2-}$$

Iodate   Sulfite   Iodide   Sulfate
ion   ion   ion   ion

When the sulfite ion is used up, the reaction is over, but the excess iodate ion reacts immediately with the iodide ion to form iodine and water.

$$IO_3{}^{1-} + 5I^{1-} + 6H^{1+} \rightarrow 3I_2 + 3H_2O$$

Iodate   Iodide   Iodine
ion   ion

A very sharp and dramatic color change occurs as the iodine forms. The appearance of the iodine can be controlled by adjusting the factors we spoke of earlier. Because of this behavior, this reaction has come to be known as the *iodine clock reaction*. (*Note:* Iodine, I_2, in dilute concentrations in water has a faint brownish color. However, if a small amount of soluble starch is added to the solution, a complex forms between the starch and the iodine, yielding an intense blue-black color. This color is very apparent even in the presence of small amounts of iodine. It is for this reason that starch is used as an indicator when performing this reaction.)

PART 1: COMPARING THE RATE OF TWO REACTIONS

PROCEDURE FOR REACTING SODIUM OXALATE AND POTASSIUM PERMANGANATE

1. Place about 15 mL of 0.1M sodium oxalate ($Na_2C_2O_4$) into a test tube and add 30 drops of 6M sulfuric acid (H_2SO_4).

2. Into a second test tube, pour enough deionized water so that it equals the volume of solution in the first test tube. This will be your control test tube.

3. Add 10 drops of 0.1M potassium permanganate ($KMnO_4$) to the control test tube and mix, using a glass stirring rod. Place this test tube in a test tube rack.

4. Have a stopwatch or a watch with a second hand ready. Add 8 drops of 0.1M potassium permanganate to the sodium oxalate solution in the first test tube, and mix, using a glass stirring rod. Place this test tube in a test tube rack and observe the change in color compared with the control test tube. Record the time it takes for the reaction to be complete. This occurs when the solution becomes colorless (not brown or tan). Also record the temperature of this solution. The reaction may take several minutes. You may proceed to the next part while you are waiting.

PROCEDURE FOR THE IODINE CLOCK REACTION

You will be working with two solutions, labeled solution A and solution B. Solution A contains 0.02M potassium iodate, and solution B contains 0.01M acidified sodium sulfate solution with starch.

1. Obtain a 400-mL beaker and place it on a piece of white paper. Label this beaker "standard run."

2. To this beaker add exactly 10 mL of solution B. Then add 80 mL of distilled water or deionized water and stir.

3. Measure out exactly 10 mL of solution A. Have your stopwatch ready and quickly pour solution A into the beaker containing solution B and the water. Stir. Measure the time interval from the point of mixing until the appearance of the blue-black color. Record this time on the report page. Also record the temperature of the solution on the report page.

PART 2: THE EFFECT OF CONCENTRATION, TEMPERATURE, AND CATALYST ON THE RATE OF REACTION

In this part of the experiment, we will examine the effect of concentration, temperature, and catalyst on the rate of the iodine clock reaction. We will compare the rate of reaction for each variable with the "standard run" performed in Part 1B.

PROCEDURE FOR SHOWING EFFECT OF CONCENTRATION ON REACTION RATE

1. Repeat steps 2 and 3 of Part 1B for each of the following solutions:

Mixture #	mL of soln A	mL of soln B	mL of water
1	10	20	70
2	10	30	60
3	20	10	70
4	30	10	60

(*Note:* Remember to first place solution B in the 400-mL beaker. Next add the water to the beaker; then add solution A.)

Record all your data on the report page.

PROCEDURE FOR SHOWING EFFECT OF TEMPERATURE ON REACTION RATE

In this part of the experiment, we are going to prepare our "standard run" mixture. However, before we mix solutions A and B, we are going to *cool* or *heat* the solutions, as described in the following steps.

1. To a 400-mL beaker, add exactly 10 mL of solution B and 80 mL of distilled or deionized water and stir. Cool this mixture to 15°C, using an ice bath.

2. Measure out exactly 10 mL of solution A and cool it to 10°C in an ice bath.

3. Pour solution A into solution B and the water as before. Measure the time interval from the point of mixing to the appearance of the blue-black color. We'll call this Mixture 5.

4. Repeat steps 1, 2, and 3 of this procedure. However, this time, warm each of the solutions to 40°C. We'll call this Mixture 6. (*Note:* The color at the completion of the reaction at this temperature will be brown rather than blue-black.)

Record your results on the report page.

PROCEDURE FOR SHOWING EFFECT OF CATALYST ON REACTION RATE

In this part of the experiment, we are going to prepare our "standard run" mixture. However, before mixing A and B, we are going to add 1 mL of $0.1M$ copper(II) sulfate ($CuSO_4$) as a catalyst.

1. To a 400-mL beaker, add exactly 10 mL of solution B, 80 mL of deionized or distilled water, and 1 mL of $0.1M$ copper(II) sulfate.

2. Measure out 10 mL of solution A and pour it into the beaker containing solution B, the water, and the copper(II) sulfate. Record the time it takes for the blue-black color to develop. We'll call this Mixture 7. Also record the temperature of the mixture on the report page.

QUESTIONS

1. How does the rate of the sodium oxalate and potassium permanganate reaction compare with the rate of the standard iodine clock reaction? (In other words, which is faster, and by how much?)

2. For the iodine clock reaction, how does the rate of the "standard run" mixture compare with the rate of:

   a) Mixture 1, in which the volume of solution B was double that in the "standard run"?

   b) Mixture 2, in which the volume of solution B was triple that in the "standard run"?

   c) Mixture 3, in which the volume of solution A was double that in the "standard run"?

   d) Mixture 4, in which the volume of solution A was triple that in the "standard run"?

3. For the iodine clock reaction, how does the rate of the "standard run" mixture compare with the rate of:

   a) Mixture 5, in which the temperature of the reactants was much cooler than in the "standard run"?

   b) Mixture 6, in which the temperature of the reactants was much warmer than in the "standard run"?

4. For the iodine clock reaction, how does the rate of the "standard run" mixture compare with the rate of Mixture 7, which contained a catalyst?

Report on Experiment 25

Name _____

Section _____ Date _____

Instructor _____

PART 1: COMPARING THE RATE OF TWO REACTIONS

	Sodium oxalate and potassium permanaganate	Iodine clock reaction (standard run)
Time reaction began		
Time reaction completed		
Time of reaction (seconds)		
Temperature of mixture		

PART 2: THE EFFECT OF CONCENTRATION, TEMPERATURE, AND CATALYST ON THE RATE OF REACTION

EFFECT OF CONCENTRATION ON REACTION RATE

Mixture number	mL of soln A	mL of soln B	mL of water	Temp. of mixture	Time of reaction, sec
1	10	20	70		
2	10	30	60		
3	20	10	70		
4	30	10	60		

EFFECT OF TEMPERATURE ON REACTION RATE

Mixture number	mL of soln A	mL of soln B	mL of water	Temp. of mixture	Time of reaction (sec)
5	10	10	80	10°C	
6	10	10	80	40°C	

EFFECT OF A CATALYST ON REACTION RATE

Mixture number	mL of soln A	mL of soln B	mL of water	Temp. of mixture	Time of reaction (sec)
7	10	10	80		

Report on Experiment 25

Name _____

Section _____ Date _____

Instructor _____

RESPONSES TO QUESTIONS

1.

2. a)

  b)

  c)

  d)

3. a)

  b)

4.

Laboratory Exercise 2
Chem-Deck© —
Writing Chemical Compounds

Time: As long as you want to keep playing

Materials: One CHEM-DECK per group

INTRODUCTION

The objective of CHEM-DECK is to give you practice in forming chemical compounds. You do this by combining cation (positive ion) cards with anion (negative ion) cards and the proper subscript (number) cards to ensure electroneutrality of the compounds. By playing card games with CHEM-DECK, you will learn the charges of the various ions, which will help you to write chemical formulas of many compounds. Of course, the element of chance is present in this game. You need the right assortment of cards to make specific compounds. However, by using your chemical knowledge, you should be able to manipulate the cards to form the *best* chemical compounds. (*Best* in this game means the compounds worth the most points; in other words, those with high molecular weights.)

Before we discuss the rules for the various games, let's review some information about CHEM-DECK. The deck consists of 88 cards. There are 18 cation (positive ion) cards, 25 anion (negative ion) cards, and 45 subscript (number) cards. You will have to cut out the cards at the end of this manual and mix them well before you begin. You will find that there are two blank cards left over, which you can keep and use as substitutes for cards that are misplaced. (You could also introduce new ions after checking with your instructor to make sure that they would form viable compounds.) Alternatively, if you have access to blank playing cards or heavy cardboard, you can make your own deck by copying the cards as they appear here.

You can play three basic card games with CHEM-DECK: rummy, five-card draw poker, and seven-card stud poker. The rules are slightly different from those of Hoyle, so we shall explain them for you. Also, you will need a periodic table of the elements for scoring, so have one available.

CHEM-DECK RUMMY

The object of the game is to obtain three compounds. It takes four cards to make a compound (anion card, cation card, and *two* subscript or number cards). The dealer gives each player 12 cards. Now you follow the regular rules of rummy as each player tries to obtain three neutral compounds.

Scoring in the game is very important and adds to the strategy of the players. The first player to obtain three compounds wins the hand and receives 50 points. This person then figures the molecular weight of each compound formed, and adds the molecular weights to get the final score. Similarly, the other players figure the molecular weights of their compounds (if they have any) and add them to get *their* final scores. The player who obtains 1,000 points wins the game.

CHEM-DECK FIVE-CARD DRAW POKER

The object of this game is to obtain *one* compound, preferably one of high molecular weight. Each player receives five cards. Players may then exchange up to four cards if they wish. The player with the compound that has the highest molecular weight wins the hand. If no player can form a compound, the player who has the heaviest ion wins the hand.

CHEM-DECK SEVEN-CARD STUD POKER

The object of this game is to obtain *one* compound. You play this game exactly like ordinary seven-card stud. Scoring is the same as in five-card draw.

EXAMPLES AND RULES FOR FORMING COMPOUNDS

You form a chemical compound from a positively charged species (cation) and a negatively charged species (anion). You must balance these two species to form an electrically neutral compound. To balance the charges, you use subscripts.

Here are suggestions to help you choose the right subscripts.

1. Write the symbol of each element of the compound, along with its oxidation number. For example, in the compound zinc chloride, write

   $Zn^{2+}Cl^{1-}$

2. You want the positive side of the compound to balance the negative side. You can do this by crisscrossing the numbers.

   $\overset{2+}{Zn}\diagdown\overset{1-}{Cl}$ or $Zn_1^{2+}Cl_2^{1-}$ or $ZnCl_2$

3. Note that you must write subscript numbers without charge (that is, without a plus or minus sign). For example, the formula for aluminum oxide is

   $\overset{3+}{Al}\diagdown\overset{2-}{O}$ or $Al_2^{3+}O_3^{2-}$ or Al_2O_3

4. Also note that subscript numbers should be in least-common-denominator form. For example, the formula for barium oxide is

   $Ba^{2+}O^{2-}$ or $Ba_2^{2+}O_2^{2-}$ or $Ba_1^{2+}O_1^{2-}$

   or BaO

Example 1. Formation of sodium carbonate

The Na^{1+} cation and the $(CO_3)^{2-}$ anion form $Na_2^{1+}(CO_3)_1^{2-}$. Or, using CHEM-DECK, we have:

Example 2. Formation of hydrogen chloride

The H^{1+} cation and the Cl^{1-} anion form $H_1^{1+}Cl_1^{1-}$. Or, using CHEM-DECK, we have:

Note that this is the only correct formula for HCl. You may not write H_2Cl_2 or H_3Cl_3 because the rules of chemical formula writing tell us that when we form a neutral compound we must use the simplest ratio of cation to anion.

SCORING: OBTAINING MOLECULAR WEIGHTS OF COMPOUNDS

Use the periodic table at the back of this laboratory manual to obtain the atomic weight of each element. Add the atomic weights of each of the elements in the compound to get the molecular weight of the compound.

Example 1. Na_2CO_3

Atomic weight of Na = 23
Atomic weight of C = 12
Atomic weight of O = 16

$$
\begin{array}{rl}
2\,Na = 2 \times 23 = & 46 \\
1\,C\ = 1 \times 12 = & 12 \\
3\,O\ = 3 \times 16 = & 48 \\
\hline
Na_2CO_3 = & 106
\end{array}
$$

Example 2. HCl   Atomic weight of H = 1
                                    Atomic weight of Cl = 35

$$1 \text{ H} = 1 \times 1 = 1$$
$$1 \text{ Cl} = 1 \times 35 = 35$$
$$\overline{\text{HCl} = 36}$$

Example 3. $Al_2(SO_4)_3$ Atomic weight of Al = 27
                                    Atomic weight of S = 32
                                    Atomic weight of O = 16

$$2 \text{ Al} = 2 \times 27 = 54$$
$$3 \text{ S} = 3 \times 32 = 96$$
$$12 \text{ O} = 12 \times 16 = 192$$
$$\overline{Al_2(SO_4)_3 = 342}$$

COMPOSITION OF THE CARD DECK

One of each of the following cards should appear in the CHEM-DECK unless otherwise designated.

Na^{1+}	Li^{1+}	Cu^{2+}	Pb^{2+}	$(C_2H_3O_2)^{1-}$	I^{1-}	$(SO_3)^{2-}$	$(HPO_4)^{2-}$
$(NH_4)^{1+}$	Cu^{1+}	Mg^{2+}	Al^{3+}	$(HCO_3)^{1-}$	Br^{1-}	$(CO_3)^{2-}$	$(CrO_4)^{2-}$
Ag^{1+}	Sn^{2+}	Hg^{2+}	H^{1-}	$(ClO_3)^{1-}$	$(HSO_4)^{1-}$	O^{2-}	$(AsO_4)^{3-}$
K^{1+}	Ca^{2+}	Ba^{2+}	$(OH)^{1-}$	$(H_2PO_4)^{1-}$	$(NO_3)^{1-}$	S^{2-}	$(PO_4)^{3-}$
H^{1+}	Zn^{2+}	Fe^{2+}	Cl^{1-}	$(NO_2)^{1-}$	F^{1-}	$(SO_4)^{2-}$	

Also, Fe^{3+}, P^{3-}, N^{3-} (two of these)

    1 (14 of these)

    2 (15 of these)

    3 (15 of these)

Supplement
*Handy Tables**

*These tables are reprinted from our text, *Basic Concepts of Chemistry*, fourth edition (Boston: Houghton Mifflin, 1988), pp. 497–505, and the end papers. Table 1 is from *CRC Handbook of Chemistry and Physics*, 1985–1986, 66th ed., copyright © The Chemical Rubber Co., CRC Press, Inc.

Table 1
Solubilities

	Acetate	Arsenate	Bromide	Carbonate	Chlorate	Chloride	Chromate	Hydroxide	Iodide	Nitrate	Oxide	Phosphate	Sulfate	Sulfide
Aluminum	W	a	W	—	W	W	—	A	W	W	a	A	W	d
Ammonium	W	W	W	W	W	W	W	W	W	W	—	W	W	W
Barium	W	w	W	w	W	W	A	W	W	W	W	A	a	d
Cadmium	W	A	W	A	W	W	A	A	W	W	A	A	W	A
Calcium	W	w	W	w	W	W	W	W	W	W	w	w	w	w
Chromium	W	—	W*	W	—	I	—	A	W	W	a	w	W†	d
Cobalt	W	A	W	A	W	W	A	A	W	W	A	A	W	A
Copper(II)	W	A	W	—	W	W	—	A	a	W	A	A	W	A
Hydrogen	W	W	W	—	W	W	—	—	W	W	W	W	W	W
Iron(II)	W	A	W	w	W	W	—	A	W	W	A	A	W	A
Iron(III)	W	A	W	—	W	W	A	A	W	W	A	w	w	d
Lead(II)	W	A	W	A	W	W	A	w	w	W	w	A	w	A
Magnesium	W	A	W	w	W	W	W	A	W	W	A	w	W	d
Mercury(I)	w	A	A	A	W	a	w	—	A	W	A	A	w	I
Mercury(II)	W	w	W	—	W	W	w	A	w	W	w	A	d	I
Nickel	W	A	W	w	W	W	A	w	W	W	A	A	W	A
Potassium	W	W	W	W	W	W	W	W	W	W	W	W	W	W
Silver	w	A	a	A	W	a	w	—	I	W	w	A	w	A
Sodium	W	W	W	W	W	W	W	W	W	W	d	W	W	W
Strontium	W	w	W	w	W	W	w	W	W	W	A	A	w	W
Tin(II)	d	—	W	—	W	W	A	A	W	d	A	A	W	A
Tin(IV)	W	—	W	—	—	W	W	w	d	—	A	—	W	A
Zinc	W	A	W	w	W	W	w	A	W	W	w	A	W	A

Abbreviations: W = soluble in water; A = insoluble in water, but soluble in acids; w = only slightly soluble in water, but soluble in acids; a = insoluble in water, and only slightly soluble in acids; I = insoluble in both water and acids; d = decomposes in water.

*$CrBr_3$

†$Cr_2(SO_4)_3$

Reprinted with permission from *CRC Handbook of Chemistry and Physics*, 66th ed. Copyright © The Chemical Rubber Co., CRC Press, Inc.

Table 2
Oxidation numbers of ions frequently used in chemistry

	1+		2+		3+
Hydrogen	H^{1+}	Calcium	Ca^{2+}	Iron(III)	Fe^{3+}
Lithium	Li^{1+}	Magnesium	Mg^{2+}	Aluminum	Al^{3+}
Sodium	Na^{1+}	Barium	Ba^{2+}		
Potassium	K^{1+}	Zinc	Zn^{2+}		
Mercury(I)*	Hg_2^{2+}*	Mercury(II)	Hg^{2+}		
Copper(I)	Cu^{1+}	Tin(II)	Sn^{2+}		
Silver	Ag^{1+}	Iron(II)	Fe^{2+}		
Ammonium	$(NH_4)^{1+}$	Lead(II)	Pb^{2+}		
		Copper(II)	Cu^{2+}		

	1−		2−		3−
Fluoride	F^{1-}	Oxide	O^{2-}	Nitride	N^{3-}
Chloride	Cl^{1-}	Sulfide	S^{2-}	Phosphide	P^{3-}
Hydroxide	$(OH)^{1-}$	Sulfite	$(SO_3)^{2-}$	Phosphate	$(PO_4)^{3-}$
Nitrite	$(NO_2)^{1-}$	Sulfate	$(SO_4)^{2-}$	Arsenate	$(AsO_4)^{3-}$
Nitrate	$(NO_3)^{1-}$	Carbonate	$(CO_3)^{2-}$		
Acetate	$(C_2H_3O_2)^{1-}$	Chromate	$(CrO_4)^{2-}$		
Chlorate	$(ClO_3)^{1-}$				

*Note that the mercury(I) ion is a diatomic ion. In other words, you never find an Hg^{1+} ion alone, but always as $Hg^{1+}-Hg^{1+}$. The two Hg^{1+} ions are bonded to each other.

Table 3
Prefixes and abbreviations

nano- = 0.000000001	nanometer = nm
micro- = 0.000001	micron = μ (Greek letter mu)
milli- = 0.001	millimeter = mm
centi- = 0.01	milliliter = mL
deci- = 0.1	milligram = mg
deca- = 10	centimeter = cm
kilo- = 1,000	centigram = cg
	decimeter = dm
	decigram = dg
	kilometer = km
	kilogram = kg

Table 4
The metric system

Length

1 millimeter = 0.001 meter = 1/1,000 meter   1 meter = 1,000 millimeters
1 centimeter = 0.01 meter = 1/100 meter   1 meter = 100 centimeters
1 decimeter = 0.1 meter = 1/10 meter   1 meter = 10 decimeters
1 kilometer = 1,000 meters   1 meter = 0.001 kilometer

Mass

1 microgram = 0.000001 gram   1 gram = 1,000,000 micrograms
1 milligram = 0.001 gram   1 gram = 1,000 milligrams
1 centigram = 0.01 gram   1 gram = 100 centigrams
1 decigram = 0.1 gram   1 gram = 10 decigrams
1 kilogram = 1,000 grams   1 gram = 0.001 kilogram

Volume

1 milliliter = 0.001 liter
1 milliliter = 1 cubic centimeter

(Cubic centimeter is abbreviated cc or cm^3.)

Table 5
Conversion of units (English-metric)

	To convert	*into*	*multiply by*
Length	inches	centimeters	2.540
	centimeters	inches	0.3937
	feet	meters	0.3048
	meters	feet	3.281
Weight	ounces	grams	28.3495
	grams	ounces	0.03527
	pounds	grams	453.5924
	grams	pounds	0.002205
Volume	liters	quarts	1.057
	quarts	liters	0.9463

Table 6
Naturally occurring isotopes of the first 15 elements

Name	Symbol	Atomic number	Mass number	Percentage natural abundance
Hydrogen-1	^1_1H	1	1	99.985
Hydrogen-2	^2_1H	1	2	0.015
Hydrogen-3	^3_1H	1	3	Negligible
Helium-3	^3_2He	2	3	0.00013
Helium-4	^4_2He	2	4	99.99987
Lithium-6	^6_3Li	3	6	7.42
Lithium-7	^7_3Li	3	7	92.58
Beryllium-9	^9_4Be	4	9	100
Boron-10	$^{10}_5B$	5	10	19.6
Boron-11	$^{11}_5B$	5	11	80.4
Carbon-12	$^{12}_6C$	6	12	98.89
Carbon-13	$^{13}_6C$	6	13	1.11
Nitrogen-14	$^{14}_7N$	7	14	99.63
Nitrogen-15	$^{15}_7N$	7	15	0.37
Oxygen-16	$^{16}_8O$	8	16	99.759
Oxygen-17	$^{17}_8O$	8	17	0.037
Oxygen-18	$^{18}_8O$	8	18	0.204
Fluorine-19	$^{19}_9F$	9	19	100
Neon-20	$^{20}_{10}Ne$	10	20	90.92
Neon-21	$^{21}_{10}Ne$	10	21	0.257
Neon-22	$^{22}_{10}Ne$	10	22	8.82
Sodium-23	$^{23}_{11}Na$	11	23	100
Magnesium-24	$^{24}_{12}Mg$	12	24	78.70
Magnesium-25	$^{25}_{12}Mg$	12	25	10.13
Magnesium-26	$^{26}_{12}Mg$	12	26	11.17
Aluminum-27	$^{27}_{13}Al$	13	27	100
Silicon-28	$^{28}_{14}Si$	14	28	92.21
Silicon-29	$^{29}_{14}Si$	14	29	4.70
Silicon-30	$^{30}_{14}Si$	14	30	3.09
Phosphorus-31	$^{31}_{15}P$	15	31	100

Table 7
Electron configurations of the elements

Atomic number	Element	1 s	2 s	2 p	3 s	3 p	3 d	4 s	4 p	4 d	4 f	5 s	5 p	5 d	5 f	6 s	6 p	6 d	6 f	7 s
1	H	1																		
2	He	2																		
3	Li	2	1																	
4	Be	2	2																	
5	B	2	2	1																
6	C	2	2	2																
7	N	2	2	3																
8	O	2	2	4																
9	F	2	2	5																
10	Ne	2	2	6																
11	Na	2	2	6	1															
12	Mg	2	2	6	2															
13	Al	2	2	6	2	1														
14	Si	2	2	6	2	2														
15	P	2	2	6	2	3														
16	S	2	2	6	2	4														
17	Cl	2	2	6	2	5														
18	Ar	2	2	6	2	6														
19	K	2	2	6	2	6		1												
20	Ca	2	2	6	2	6		2												
21	Sc	2	2	6	2	6	1	2												
22	Ti	2	2	6	2	6	2	2												
23	V	2	2	6	2	6	3	2												
24	Cr	2	2	6	2	6	5	1												
25	Mn	2	2	6	2	6	5	2												
26	Fe	2	2	6	2	6	6	2												
27	Co	2	2	6	2	6	7	2												
28	Ni	2	2	6	2	6	8	2												
29	Cu	2	2	6	2	6	10	1												
30	Zn	2	2	6	2	6	10	2												
31	Ga	2	2	6	2	6	10	2	1											
32	Ge	2	2	6	2	6	10	2	2											
33	As	2	2	6	2	6	10	2	3											
34	Se	2	2	6	2	6	10	2	4											
35	Br	2	2	6	2	6	10	2	5											
36	Kr	2	2	6	2	6	10	2	6											
37	Rb	2	2	6	2	6	10	2	6			1								
38	Sr	2	2	6	2	6	10	2	6			2								
39	Y	2	2	6	2	6	10	2	6	1		2								
40	Zr	2	2	6	2	6	10	2	6	2		2								
41	Nb	2	2	6	2	6	10	2	6	4		1								
42	Mo	2	2	6	2	6	10	2	6	5		1								
43	Tc	2	2	6	2	6	10	2	6	6		1								
44	Ru	2	2	6	2	6	10	2	6	7		1								
45	Rh	2	2	6	2	6	10	2	6	8		1								
46	Pd	2	2	6	2	6	10	2	6	10										
47	Ag	2	2	6	2	6	10	2	6	10		1								
48	Cd	2	2	6	2	6	10	2	6	10		2								
49	In	2	2	6	2	6	10	2	6	10		2	1							
50	Sn	2	2	6	2	6	10	2	6	10		2	2							
51	Sb	2	2	6	2	6	10	2	6	10		2	3							
52	Te	2	2	6	2	6	10	2	6	10		2	4							
53	I	2	2	6	2	6	10	2	6	10		2	5							
54	Xe	2	2	6	2	6	10	2	6	10		2	6							
55	Cs	2	2	6	2	6	10	2	6	10		2	6			1				
56	Ba	2	2	6	2	6	10	2	6	10		2	6			2				
57	La	2	2	6	2	6	10	2	6	10		2	6	1		2				
58	Ce	2	2	6	2	6	10	2	6	10	2	2	6			2				
59	Pr	2	2	6	2	6	10	2	6	10	3	2	6			2				
60	Nd	2	2	6	2	6	10	2	6	10	4	2	6			2				

Table 7
Electron configurations of the elements (Continued)

Atomic number	Element	1	2		3			4				5				6				7
		s	s	p	s	p	d	s	p	d	f	s	p	d	f	s	p	d	f	s
61	Pm	2	2	6	2	6	10	2	6	10	5	2	6			2				
62	Sm	2	2	6	2	6	10	2	6	10	6	2	6			2				
63	Eu	2	2	6	2	6	10	2	6	10	7	2	6			2				
64	Gd	2	2	6	2	6	10	2	6	10	7	2	6	1		2				
65	Tb	2	2	6	2	6	10	2	6	10	9	2	6			2				
66	Dy	2	2	6	2	6	10	2	6	10	10	2	6			2				
67	Ho	2	2	6	2	6	10	2	6	10	11	2	6			2				
68	Er	2	2	6	2	6	10	2	6	10	12	2	6			2				
69	Tm	2	2	6	2	6	10	2	6	10	13	2	6			2				
70	Yb	2	2	6	2	6	10	2	6	10	14	2	6			2				
71	Lu	2	2	6	2	6	10	2	6	10	14	2	6	1		2				
72	Hf	2	2	6	2	6	10	2	6	10	14	2	6	2		2				
73	Ta	2	2	6	2	6	10	2	6	10	14	2	6	3		2				
74	W	2	2	6	2	6	10	2	6	10	14	2	6	4		2				
75	Re	2	2	6	2	6	10	2	6	10	14	2	6	5		2				
76	Os	2	2	6	2	6	10	2	6	10	14	2	6	6		2				
77	Ir	2	2	6	2	6	10	2	6	10	14	2	6	7		2				
78	Pt	2	2	6	2	6	10	2	6	10	14	2	6	9		1				
79	Au	2	2	6	2	6	10	2	6	10	14	2	6	10		1				
80	Hg	2	2	6	2	6	10	2	6	10	14	2	6	10		2				
81	Tl	2	2	6	2	6	10	2	6	10	14	2	6	10		2	1			
82	Pb	2	2	6	2	6	10	2	6	10	14	2	6	10		2	2			
83	Bi	2	2	6	2	6	10	2	6	10	14	2	6	10		2	3			
84	Po	2	2	6	2	6	10	2	6	10	14	2	6	10		2	4			
85	At	2	2	6	2	6	10	2	6	10	14	2	6	10		2	5			
86	Rn	2	2	6	2	6	10	2	6	10	14	2	6	10		2	6			
87	Fr	2	2	6	2	6	10	2	6	10	14	2	6	10		2	6			1
88	Ra	2	2	6	2	6	10	2	6	10	14	2	6	10		2	6			2
89	Ac	2	2	6	2	6	10	2	6	10	14	2	6	10		2	6	1		2
90	Th	2	2	6	2	6	10	2	6	10	14	2	6	10		2	6	2		2
91	Pa	2	2	6	2	6	10	2	6	10	14	2	6	10	2	2	6	1		2
92	U	2	2	6	2	6	10	2	6	10	14	2	6	10	3	2	6	1		2
93	Np	2	2	6	2	6	10	2	6	10	14	2	6	10	4	2	6	1		2
94	Pu	2	2	6	2	6	10	2	6	10	14	2	6	10	6	2	6			2
95	Am	2	2	6	2	6	10	2	6	10	14	2	6	10	7	2	6			2
96	Cm	2	2	6	2	6	10	2	6	10	14	2	6	10	7	2	6	1		2
97	Bk	2	2	6	2	6	10	2	6	10	14	2	6	10	8	2	6	1		2
98	Cf	2	2	6	2	6	10	2	6	10	14	2	6	10	10	2	6			2
99	Es	2	2	6	2	6	10	2	6	10	14	2	6	10	11	2	6			2
100	Fm	2	2	6	2	6	10	2	6	10	14	2	6	10	12	2	6			2
101	Md	2	2	6	2	6	10	2	6	10	14	2	6	10	13	2	6			2
102	No	2	2	6	2	6	10	2	6	10	14	2	6	10	14	2	6			2
103	Lr	2	2	6	2	6	10	2	6	10	14	2	6	10	14	2	6	1		2
104	Unq	2	2	6	2	6	10	2	6	10	14	2	6	10	14	2	6	2		2
105	Unp	2	2	6	2	6	10	2	6	10	14	2	6	10	14	2	6	3		2
106	Unh	2	2	6	2	6	10	2	6	10	14	2	6	10	14	2	6	4		2
107	Uns	2	2	6	2	6	10	2	6	10	14	2	6	10	14	2	6	5		2
108	Uno	2	2	6	2	6	10	2	6	10	14	2	6	10	14	2	6	6		2
109	Une	2	2	6	2	6	10	2	6	10	14	2	6	10	14	2	6	7		2

Table 8
Heats of formation at 25°C and 1 atmosphere pressure

Substance	ΔH_f (kcal/mole)
$BaCl_2$ (s)	−205.56
BaO_2 (s)	−150.5
CH_4 (g)	−17.9
C_2H_2 (g)	+54.2
C_2H_6 (g)	−20.2
C_8H_{18} (l)	−59.97
$C_8H_{12}O_6$ (s)	−304.4
CO (g)	−26.4
CO_2 (g)	−94.1
CaO (s)	−151.9
$CaCO_3$ (s)	−288.5
HI (g)	+6.2
HCl (g)	−22.1
H_2O (g)	−57.8
H_2O (l)	−68.3
H_2O_2 (l)	−44.8
H_2S (g)	−4.8
H_3PO_4 (aq)	−308.2
SO_2 (g)	−70.96
SO_3 (g)	−94.45

Table 9
Pressure of water vapor, P_{H_2O}, at various temperatures

Temperature (°C)	Pressure (torr)	Temperature (°C)	Pressure (torr)
0	4.58	32	35.66
5	6.54	33	37.73
10	9.21	34	39.90
15	12.79	35	42.18
16	13.63	36	44.56
17	14.53	37	47.07
18	15.48	38	49.69
19	16.48	39	52.44
20	17.54	40	55.32
21	18.65	45	71.88
22	19.83	50	92.51
23	21.07	55	118.04
24	22.38	60	149.38
25	23.76	65	187.54
26	25.21	70	233.7
27	26.74	75	289.1
28	28.35	80	355.1
29	30.04	85	433.6
30	31.82	90	525.8
31	33.70	95	633.9
		100	760.0

Table 10
Alphabetical list of the elements

Name of element	Symbol	Atomic number	Atomic weight
Actinium	Ac	89	(227)
Aluminum	Al	13	26.9815
Americium	Am	95	(243)
Antimony	Sb	51	121.75
Argon	Ar	18	39.948
Arsenic	As	33	74.9216
Astatine	At	85	(210)
Barium	Ba	56	137.34
Berkelium	Bk	97	(247)
Beryllium	Be	4	9.0122
Bismuth	Bi	83	208.980
Boron	B	5	10.811
Bromine	Br	35	79.904
Cadmium	Cd	48	112.40
Calcium	Ca	20	40.08
Californium	Cf	98	(251)
Carbon	C	6	12.01115
Cerium	Ce	58	140.12
Cesium	Cs	55	132.905
Chlorine	Cl	17	35.453
Chromium	Cr	24	51.996
Cobalt	Co	27	58.9332
Copper	Cu	29	63.546
Curium	Cm	96	(247)
Dysprosium	Dy	66	162.50
Einsteinium	Es	99	(254)
Erbium	Er	68	167.26
Europium	Eu	63	151.96
Fermium	Fm	100	253
Fluorine	F	9	18.9984
Francium	Fr	87	(223)
Gadolinium	Gd	64	157.25
Gallium	Ga	31	69.72
Germanium	Ge	32	72.59
Gold	Au	79	196.967
Hafnium	Hf	72	178.49
Helium	He	2	4.0026
Holmium	Ho	67	164.930
Hydrogen	H	1	1.00797
Indium	In	49	114.82
Iodine	I	53	126.9044
Iridium	Ir	77	192.2
Iron	Fe	26	55.847
Krypton	Kr	36	83.80
Lanthanum	La	57	138.91
Lawrencium	Lr	103	(257)
Lead	Pb	82	207.19
Lithium	Li	3	6.939
Lutetium	Lu	71	174.97
Magnesium	Mg	12	24.312
Manganese	Mn	25	54.9380
Mendelevium	Md	101	(256)
Mercury	Hg	80	200.59
Molybdenum	Mo	42	95.94
Neodymium	Nd	60	144.24
Neon	Ne	10	20.183
Neptunium	Np	93	(237)
Nickel	Ni	28	58.71
Niobium	Nb	41	92.906
Nitrogen	N	7	14.0067
Nobelium	No	102	(254)

Table 10
Alphabetical list of the elements (Continued)

Name of element	Symbol	Atomic number	Atomic weight
Osmium	Os	76	190.2
Oxygen	O	8	15.9994
Palladium	Pd	46	106.4
Phosphorus	P	15	30.9738
Platinum	Pt	78	195.09
Plutonium	Pu	94	(242)
Polonium	Po	84	(210)
Potassium	K	19	39.102
Praseodymium	Pr	59	140.907
Promethium	Pm	61	(147)
Protactinium	Pa	91	(231)
Radium	Ra	88	(226)
Radon	Rn	86	(222)
Rhenium	Re	75	186.2
Rhodium	Rh	45	102.905
Rubidium	Rb	37	85.47
Ruthenium	Ru	44	101.07
Samarium	Sm	62	150.35
Scandium	Sc	21	44.956
Selenium	Se	34	78.96
Silicon	Si	14	28.086
Silver	Ag	47	107.868
Sodium	Na	11	22.9898
Strontium	Sr	38	87.62
Sulfur	S	16	32.064
Tantalum	Ta	73	180.948
Technetium	Tc	43	(99)
Tellurium	Te	52	127.60
Terbium	Tb	65	158.924
Thallium	Tl	81	204.37
Thorium	Th	90	232.038
Thulium	Tm	69	168.934
Tin	Sn	50	118.69
Titanium	Ti	22	47.90
Tungsten	W	74	183.85
Unnilenium	Une	109	(267)
Unnihexium	Unh	106	(263)
Unniloctium	Uno	108	(265)
Unnilpentium	Unp	105	(262)
Unnilquadium	Unq	104	(261)
Unnilseptium	Uns	107	(262)
Uranium	U	92	238.03
Vanadium	V	23	50.942
Xenon	Xe	54	131.30
Ytterbium	Yb	70	173.04
Yttrium	Y	39	88.905
Zinc	Zn	30	65.37
Zirconium	Zr	40	91.22

Note: Values in parentheses are atomic weights of isotopes with longest half-life, except for Pm, Po, and Tc, whose best-known isotopes are listed. Atomic weights are based on ^{12}C.

Alphabetical Listing of Chemicals

The following is an alphabetical listing of the chemicals used in this lab manual. The numbers following each chemical indicate the experiments in which the chemical is used.

acetic acid (or commercial vinegar), 12, 14, 21
acetone, 21
albumin, 22
alum (potassium aluminum sulfate dodecahydrate), 13
aluminum nitrate, 3
ammonium chloride, 12, 15
ammonium hydroxide, 3, 12, 15
ammonium molybdate, 11
amyl acetate, 21
arsenic pentoxide, 8
ascorbic acid, 11

barium chloride dihydrate, 3, 6, 13
Benedict's solution, 22
benzoic acid, 17
biuret solution, 22
butyric acid, 7

calcium carbonate, 15
calcium chloride, 6, 15
calcium oxide, 8
calcium sulfate dihydrate, 13
carbon chips, 8
chloroform, 17
citric acid, 17
cobalt(II) chloride, 4
copper strips, 19
copper sulfate pentahydrate, 13, 19, 22, 25
copper turnings, 8
copper wire, 19
cottonseed oil, 22
cyclohexane, 21
cyclopentane, 21

Ca^{+2} | Sn^{+2} | Cu^{+1}
Calcium | Tin (II) | Copper (I)

Calcium
Ca^{+2}
Tin (II)
Sn^{+2}
Copper (I)
Cu^{+1}

Li^{+1} H^{+1} K^{+1}
Lithium   Hydrogen   Potassium

Lithium
Li^{+1}
Hydrogen
H^{+1}
Potassium
K^{+1}

Ag^{+1} (NH$_4$)$^{+1}$ Na^{+1}
Silver   Ammonium   Sodium

Silver
Ag^{+1}
Ammonium
(NH$_4$)$^{+1}$
Sodium
Na^{+1}

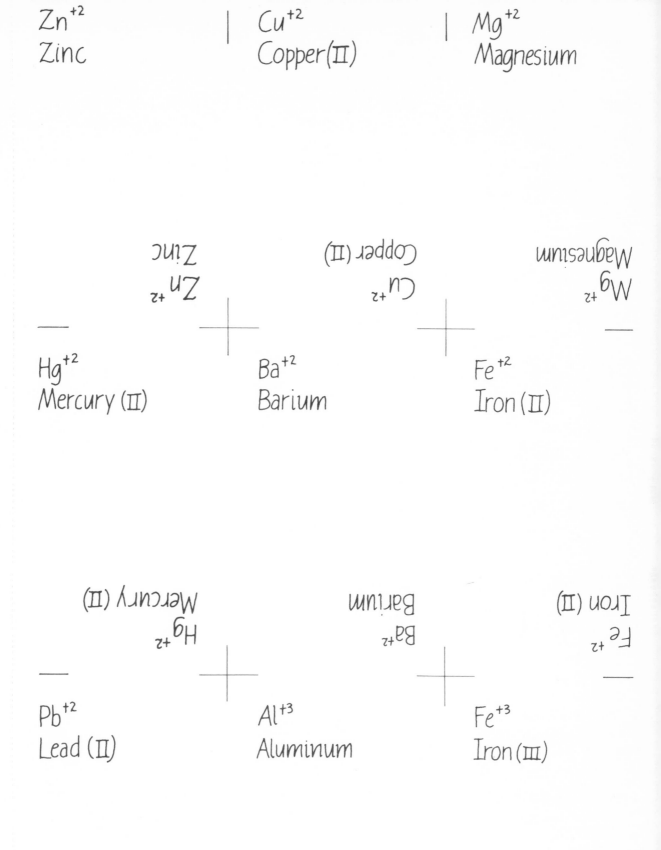

Zn⁺² Zinc | Cu⁺² Copper(II) | Mg⁺² Magnesium

Zn⁺² Zinc | Cu⁺² Copper(II) | Mg⁺² Magnesium

Hg⁺² Mercury(II) | Ba⁺² Barium | Fe⁺² Iron(II)

Hg⁺² Mercury(II) | Ba⁺² Barium | Fe⁺² Iron(II)

Pb⁺² Lead(II) | Al⁺³ Aluminum | Fe⁺³ Iron(III)

(Inverted:)

Pb⁺² Lead(II) | Al⁺³ Aluminum | Fe⁺³ Iron(III)

Fe^{+3}
Iron (III)

H^{-1}
Hydride

$(OH)^{-1}$
Hydroxide

Iron (III)
Fe^{+3}

Hydride
H^{-1}

Hydroxide
$(OH)^{-1}$

Cl^{-1}
Chloride

$(C_2H_3O_2)^{-1}$
Acetate

$(HCO_3)^{-1}$
Hydrogen carbonate

Chloride
Cl^{-1}

Acetate
$(C_2H_3O_2)^{-1}$

Hydrogen carbonate
$(HCO_3)^{-1}$

$(ClO_3)^{-1}$
Chlorate

$(H_2PO_4)^{-1}$
Dihydrogen phosphate

$(NO_2)^{-1}$
Nitrite

Chlorate
$(ClO_3)^{-1}$

Dihydrogen phosphate
$(H_2PO_4)^{-1}$

Nitrite
$(NO_2)^{-1}$

I^{-1}
Iodide

Br^{-1}
Bromide

$(HSO_4)^{-1}$
Hydrogen
sulfate

Iodide
I^{-1}

Bromide
Br^{-1}

Hydrogen
sulfate
$(HSO_4)^{-1}$

$(NO_3)^{-1}$
Nitrate

F^{-1}
Fluoride

$(SO_3)^{-2}$
Sulfite

Nitrate
$(NO_3)^{-1}$

Fluoride
F^{-1}

Sulfite
$(SO_3)^{-2}$

$(CO_3)^{-2}$
Carbonate

O^{-2}
Oxide

S^{-2}
Sulfide

Carbonate
$(CO_3)^{-2}$

Oxide
O^{-2}

Sulfide
S^{-2}

$(SO_4)^{-2}$
Sulfate | $(HPO_4)^{-2}$
Hydrogen
phosphate | $(CrO_4)^{-2}$
Chromate

Sulfate
$(SO_4)^{-2}$ | Hydrogen
phosphate
$(HPO_4)^{-2}$ | Chromate
$(CrO_4)^{-2}$

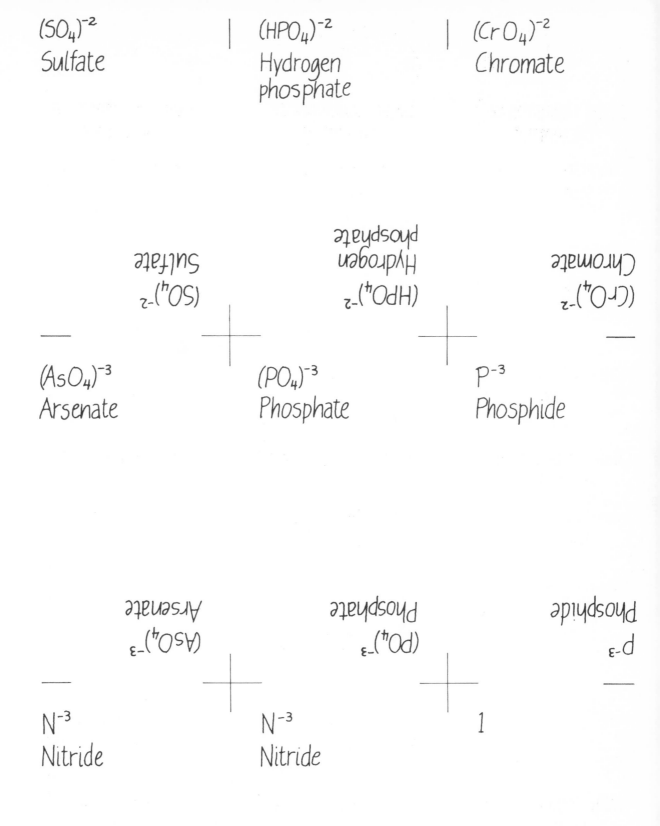

$(AsO_4)^{-3}$
Arsenate | $(PO_4)^{-3}$
Phosphate | P^{-3}
Phosphide

Arsenate
$(AsO_4)^{-3}$ | Phosphate
$(PO_4)^{-3}$ | Phosphide
P^{-3}

N^{-3}
Nitride | N^{-3}
Nitride | 1

Nitride
N^{-3} | Nitride
N^{-3} | 1

$$\frac{1}{1} \qquad +\frac{1}{2} \qquad +\frac{1}{2} \qquad \frac{1}{1}$$

$$\frac{1}{2} \qquad +\frac{1}{2} \qquad +\frac{2}{2} \qquad \frac{2}{2}$$

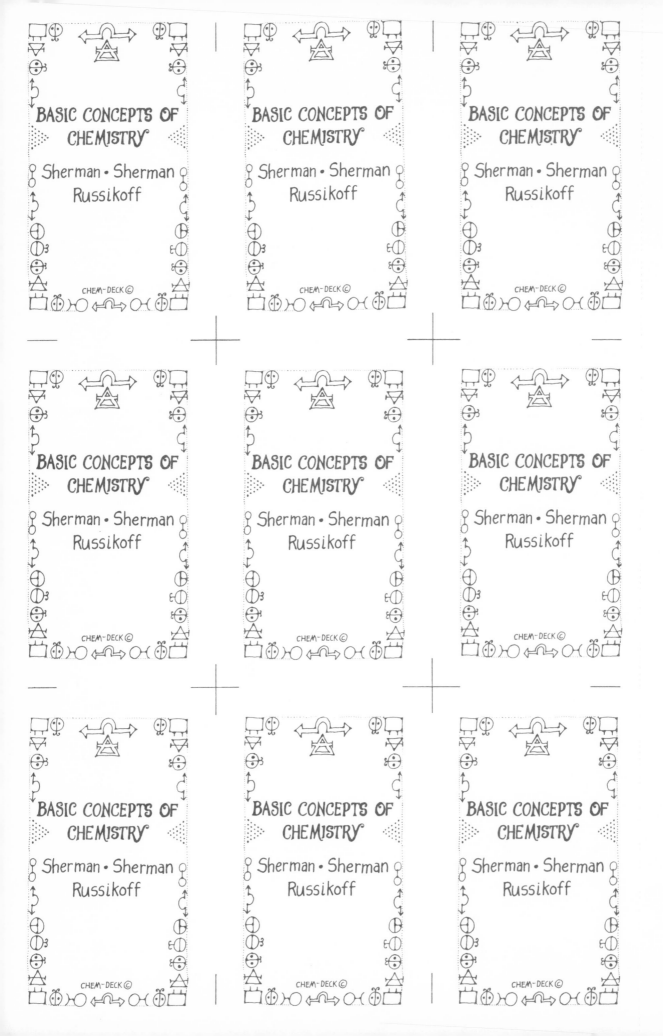

2

2

2

2

$\dfrac{}{2}$

$+\dfrac{}{2}$

$+\dfrac{}{2}$

$\dfrac{}{2}$

$\dfrac{}{2}$

$+\dfrac{}{2}$

$+\dfrac{}{2}$

$\dfrac{}{2}$

2

2

2

2

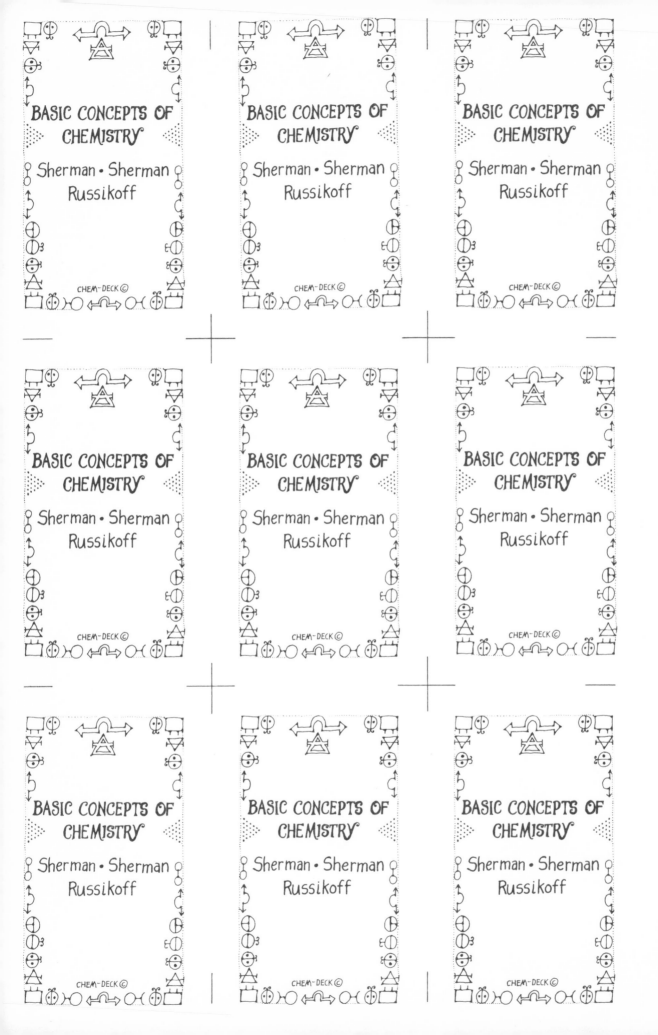

$\dfrac{2}{}$ $\dfrac{}{3}$ $\dfrac{}{3}$

$\dfrac{}{3}$ $+\dfrac{2}{3}$ $+\dfrac{3}{3}$ $\dfrac{3}{}$

$\dfrac{}{3}$ $+\dfrac{3}{3}$ $+\dfrac{3}{3}$ $\dfrac{3}{}$

$\dfrac{}{3}$ $\dfrac{}{3}$ $\dfrac{}{3}$

3 | 3 | 3

3
—
3

$\overline{\quad}$ 3

+ 3

3 +

+ 3

3 +

$\overline{\quad}$ 3

3
—
3

3 +

+

+

3 +

$\overline{\quad}$ 3

3 |

|